Eel Culture

Eel Culture

by Atsushi Usui

Shizuoka Tansui Company
(Freshwater Fish Farming Company)
Shizuoka Prefecture

Translated by Ichiro Hayashi

Tokai Regional Fisheries Research Laboratory

Much co-ordination and help in re-arranging
and supplementing the material for western
needs and world audience was given by
Dr Gordon Williamson who had himself studied
in Japan at Shimonoseki University of Fisheries.

Fishing News (Books) Ltd
23 Rosemount Avenue, West Byfleet, Surrey, England
and 110 Fleet Street, London EC4A 2JL

Tokai Regional Fisheries
Research Laboratory
Tokyo, Japan

ISBN 0 85238 024 0

design/computer composition/printed in England by
Eyre & Spottiswoode Ltd at Grosvenor Press

Contents

5

List of Illustrations

[Beginning page 127]

6

7

List of Tables

9

Preface

The basic manuscript of this book was first prepared in Japanese by Mr Usui, a distinguished biologist and specialist in the technique of eel culture. In the preface to that work Usui wrote:-

'At first it was natural for me, reflecting on my lack of ability, to refuse requests to write a book about eel culture, as I felt extremely uneasy of success in compiling the material for a useful book.

'However, encouraged by many strong requests, I made up my mind to do my best to complete a book containing my own professional viewpoints that could be used by beginners in their own studies.

'During the war eel culture in Japan was severely handicapped by the fact that most ponds were converted into rice fields in order to meet the basic needs of human nutrition and there was a great shortage of feed for eels (eels require seven times their body weight in feed). Now, Japan is recovered from those bad conditions. The times have so prospered that despite a notably large quantity of eels being cultured the increasing demand makes it impossible to maintain balance between consumption and production.

'I am concerned that eel culture has largely been developed without a detailed understanding of disease and nutrition problems.

'Experts have devoted their studies almost entirely to the development and commercial profit of the industry without undertaking scientific research, and to my surprise I have encountered some serious difficulty in making a study of eel culture. Therefore, those who plan to begin eel culture should take the above-mentioned factors into consideration.

'However, from a commercial point of view, it often happens that unexpected benefits develop regardless of unsettled problems thus turning a misfortune into a blessing for industry.

'It is my great pleasure as a writer thus to have been able to set down my opinions in spite of the impossibility of fully covering all the facets of this subject. In this book, working from my own experience, I have made quotations of accurate statistics as far as the conditions permit, so that the reader

11

may learn about eel culture for his own knowledge and practical use.

'I should like to express my great appreciation both to Dr. Densaburo Inaba, ex-Professor of the Tokyo University of Fisheries, for his kind supervision, and to Dr. Michiaki Ochi, ex-Dean of the Fisheries Department of the Kagoshima University, for his kind revision.'

This manuscript was translated into English by Ichiro Hayashi and submitted to Fishing News (Books) Ltd. Its value was immediately apparent as being the first detailed account of the methods and techniques of eel culture that have evolved in Japan in recent decades. With the permission (courteously and readily granted) of the author, it was arranged in order to make the manuscript fully useful and acceptable to those it was intended to help, that the material should be supplemented and adapted into customary English format.

This work was undertaken by Dr. Gordon Williamson, a British biologist of experience, who had himself recently been in Japan for nearly a year engaging in the study of eels. In his period in Japan Dr. Williamson secured and perused a number of Japanese books on the subject and after perusal of Mr. Usui's work he affirmed it was the best he had encountered.

While he was in Japan, Dr. Williamson was able to study (by invitation) under Dr. Isao Matsui, president of Shimonoseki University of Fisheries, to whom he gratefully extends his thanks and acknowledgement of benefits gained and knowledge acquired by that study.

To the translation, Dr. Williamson added certain new material in order to make the work fully comprehensive and acceptable to the world market and give all possible information to readers in other countries. This additional material and the re-arrangement of chapters which was undertaken for logical reasons, has added much to the value of the work (while considerably increasing its extent) so that now it is presented to the world market as the first comprehensive work on the culture of eels in a style suitable to readers of the English language. Line drawings and sketches have necessarily been redrawn and a number of new photographs added to the excellent and comprehensive collection originally supplied from Japan. The existing methods of eel culture are described in detail together with an estimate of possible future developments and much added statistical material of importance.

The book is now commended to readers as the combined result of Japanese technical skill portrayed by Mr Usui and Dr. Williamson's comprehensive biological knowledge and literary craftsmanship, both made possible initially by Ichiro Hayashi's able translation. One outstanding additional item is worthy of particular mention. This is a reproduction of the original photograph published in 1909 by Dr. Johannes Schmidt of the progressive stages of the leptocephalus in its metamorphosis into the elver form of the eel. It was that photograph and the accompanying report that announced the discovery that the European eel did not spawn in freshwater but in the Sargasso sea area of the Atlantic ocean. This is the first time that this original photograph has been made available since 1909 for publication and thanks are given to the Danish Carlsberg Foundation concerned for the privilege of using it in this book.

Fig 1   A group of eel culture ponds, in Yoshida District, Shizuoka Prefecture, the most famous intensive farming eel culture district in Japan. (The River Ohi is seen in the upper

Publishers Note

The illustrations and supporting captions constitute a most valuable part of this work. For technical reasons in production and the need for varied cross references it has not been possible to incorporate the illustrations satisfactorily into the text; therefore the attention of readers is specifically directed to the need for paying special attention to the illustrations, the captions to which have accordingly been made as ample and informative as possible.

Certain pages have been left blank at the end of chapters for the convenience of operators to make additional notes as gained by experience.

Conversion Factors

The conversion factors used in the book are:
1 hectare (ha) = 10,000 m^2 = 2.5 acres
1 m^3 = 220 gallons
1 metric ton = 1000 kg = 2204 pounds

1. Introduction

Eel culture was started in Japan from the year 1894 and now the huge eel culture industry of that country produces about 24,000 tons of eels per year and is still growing. Taiwan and Korea have copied Japan's methods and in Europe and in Australia and New Zealand there is much interest in getting eel culture started.

We must note that in a protein-hungry world it is highly wasteful to culture carnivorous animals such as eels, but that is the way it goes. An eel must eat 7 kg of fresh or 1.3-1.5 kg of the commercial diet (composed mainly of white fish meal[1] *; in order to gain 1 kg body weight itself.

The Japanese eel species is called [Anguilla japonica] or [unagi] in Japanese and is one of 16 species of [Anguilla] eels in the world. All these eels have to a great extent similar habits and can undoubtedly be cultured by the same method as is used in Japan. The appearance of the Japanese eel [A.japonica], European eel [A.anguilla], American eel [A.rostrata] and Australian and New Zealand eel [A.australis] are so alike that they are identical from a marketing point of view.

Table 1 shows the weights of eels caught and eaten in various areas of the world. About 25,000 tons of wild eels and 26,000 tons of cultured eels are harvested in the world in average years. The main eel-eating areas are Europe and the Far East: in many other areas of the world where eels occur they are not eaten because people are afraid of their snake-like appearance. The main catches of eels are made in Europe, North America, the Far East, Australia and New Zealand.

The bodies of water in the world giving the heaviest sustained catches of wild eels are given in Table 2. Although many [Anguilla] eel species occur in the tropics, actual stocks of eels and catches are only small in tropical areas.

The flesh of eels consists of delicious sweet meat with only one bone down the middle and there are at least three styles of cooking:

* smoked — the method used on the continent of Europe
* jellied or stewed — the London method
* kabayaki — the Japanese style

Almost every eel-producing area of the world is now exploited to yield the thousands of tons of eels eaten in Europe each year. Tanker-lorries bring live wild-caught eels from all over Europe, N. Africa and N. America while frozen wild eels are imported from New Zealand and Australia.[2]

The demand increases annually — as does the price! But there are no more unexploited wild stocks of eels in existence. Therefore the only way to obtain further supplies of this delicious food is to culture eels and the purpose of this book is to outline the practical methods that have evolved in Japan, in particular, for achieving that increased supply.

[1] Peru's fish meal is composed of anchovy which is *brown* fish meal. We in Japan are therefore using white fish meal only, because of realizing experimental results of higher efficiency in culturing eels.
[2] Japan has not exported frozen cultured eels since 1968, because the local demand has increased with a corresponding increase in price.

2. The world's species of *Anguilla* eels

The eels discussed in this book are members of a very unusual tropical genus of fish called [Anguilla], which spend part of their lives in the sea and part in fresh water. There are thousands of eel species in the world that spend their whole lives in the sea and hundreds of species which live their whole lives in freshwater. But by a quirk of nature, only the [Anguilla] eels have a life history half in the sea and half in freshwater.

The peculiar habits of [Anguilla] eels have mystified men for centuries. No-one has ever found ripe eggs in [Anguilla] eels. All other freshwater fish contain ripe eggs at some season, but never eels. How then do eels reproduce? Aristotle reckoned eels were created "in the bowels of the earth". English country folk believed elvers were fallen horse hairs that came to life when dropped in a puddle. Ranging from Europe to far Tahiti, magic origins were attributed to eels.

In fact, eels breed in the sea, far from land, and their sexual organs do not mature until after the adult eels migrate back from the rivers into the sea. That is why no eggs or sperm can ever be found in an eel captured in freshwater. Many ages ago [Anguilla] eels were marine fish like Conger eels, and later they somehow developed the habit of entering freshwater during the growing period of their lives.

The story of how the amazing life cycle of [Anguilla] eels was discovered is a major epic of science and to this day important aspects of the lives of eels still remain to be elucidated. The early scientific investigations on eels all centred on the European species [Anguilla anguilla]. This was due to the fact that in the nineteenth century scientific research existed only in Europe.

19

Two Italians, Grassi and Calandruccio, made the first important discovery in 1897 when they kept alive in a sea aquarium some peculiar transparent leaf-shaped fishes called [Leptocephalus] caught in the sea near Messina. To their amazement they watched these fishes during two months change their shape to become elvers of the freshwater eel (Figs 2 and 4). This showed that eels bred in the sea, and that the fish previously called a [Leptocephalus] was actually the larval stage of an eel. Grassi and Calandruccio suggested that the spawning place of eels was near Messina in the Mediterranean.

But they were wrong. The eel's amazing life migration remained a mystery for the Danish biologist Johannes Schmidt to elucidate.

As a young man then studying cod, Schmidt in 1904 was making plankton hauls to catch cod eggs far out at sea in the Atlantic off the Faroes when he found in his net - a leptocephalus larva of an eel! Schmidt immediately realised that he had discovered a vital clue to the life history of the eel. So, he thought, eel larvae occurred at Messina and off Faroe - where else? That one larva started him on a research mission that lasted 35 years. With the backing of the Danish Government and the Carlsberg Foundation, Schmidt gradually made cruises and plankton hauls all down the coasts of Europe. He caught many more eel larvae - all large ones. So he wondered, in what part of the ocean did small larvae occur?

After the First World War, Schmidt was able to make several cruises to the central and eastern parts of the Atlantic and in 1922 he announced the famous result of his persistent research. It was that the central part of the Atlantic called the Sargasso Sea was the only area where newly-hatched tiny eel larvae occurred (Fig 3). The extraordinary European eel was thus shown to spawn in only one place - in the centre of the ocean far from land and in the tropical zone of the world. The ancestors of [Anguilla] eels were tropical marine eels which lived their whole life cycle in the sea.

Spawning takes place in February each year and is thought to occur about 400 metres beneath the surface, in water of about 17°C temperature (the eels spawn in *mid-water*, not on the bottom). After being liberated by the adult eels the eggs rise and float near the surface where they hatch in about 24

20

hours into tiny prelarvae 5 mm long. These tiny planktonic fish gradually grow into transparent, leaf-shaped leptocephalae larvae and get carried away from the Sargasso area by the Gulf Stream which flows from that area away to the north east. After drifting for about 22 months the leptocephalae arrive over the Continental shelf of Europe (which begins far from the actual coast) in November each year and there metamorphose into slim elvers. Possibly attracted by the smell of fresh water or some other driving force these transparent elvers in millions head for the coast and enter river mouths. Because of temperature factors and the rate of current flow in different areas their arrival varies at different parts of the European coast. In the rivers of Northern Ireland, the southern and western rivers of England and Scotland and of Belgium, Denmark and Holland, the main arrival of elvers occurs in the months of April and May. Their size at the end of this journey is such that on average some 3,500 are required to weigh a kilogramme.

Further south in the warmer waters of the French coast the arrival begins as early as February and the size is larger, needing only 2,800 to the kilogramme. Further south still on the coasts of Portugal and Spain the run begins as early as December and January and again, being fractionally larger because of having lost less weight they go 2,700 to the kilogramme.

The American eel [A.rostrata] also breeds in the Sargasso area and its elvers take about ten months to reach the American continental shelf.

In freshwater, the elvers turn black in colour and courageously migrate inland, nosing against the current at every stage, climbing around waterfalls by wriggling up the mossy sides of the falls (Fig 5). They feed actively, mainly on insects and other small animals and are wholly carnivorous; their diet is closely similar to that of trout. Although the great majority of eels spend their growing lives in freshwater, a minority stay in salt or brackish water coastal localities. Eels can be transferred from sea to freshwater and vice versa without any harm being caused.

When they are finally fully grown after various lengthy periods, the eels migrate downstream into the sea in autumn and disappear from man's knowledge. Somehow, at least some of them must successfully navigate their way to the

Sargasso spawning area. But no-one has yet caught a single adult eel under the high seas. Nevertheless by deduction they reach the Sargasso and spawn. Each female lays from 7-13 million eggs according to her size. Since no big eels ever re-enter rivers, we know that the adults must die after spawning. The ocean bottom of the Sargasso must be covered with their bones.

DIFFERENCES AMONG THE WORLD'S EEL SPECIES

His discovery made Schmidt famous for life. But he wanted further discoveries and so he immediately set his student Vilhelm Ege to try to discover how many other species of [Anguilla] existed in the world.

From museums he knew that many [Anguilla] eels existed in the Indian and Pacific oceans. In fact, over-enthusiastic biologists had described over 100 species. By years of patient correspondence with museums, Danish consuls, missionaries, ships captains and business men, persuading them to send collections of eels from every part of the world, and finally going themselves on a round-the-world eel-collecting expedition on the research ship DANA, Schmidt and Ege accumulated in Copenhagen a mammoth collection of 12,793 adult eels and 12,472 elvers from all over the world.

This vast collection was next examined in detail. Schmidt died in 1933 but Ege carried on. Finally in 1939, Ege published the results that everyone had been waiting for. Truly both men gave their lives to eels.

Ege's list showed sixteen species of eels in the world and that the homeland of the Anguilla eels in which they have evolved and from whence they have spread to other regions is the Indonesian Archipelago (Fig 6).

The spawning places for South Pacific and Indian Ocean eels were investigated and believed determined by Brunn (1937), Ege (1939) and Jespersen (1942). According to their studies the probable area for spawning was in the sea territory about 400 metres depth, above 13°C in temperature and above 3.4 per cent salinity ranging from the Bonin Islands to the Okinawa coast (far south of Japan). The collecting records of specimens of leptocephalus were as follows: 19 specimens at about 120 miles southeast of Okinawa Island in 1961 and one specimen at about 30 miles south of the southern edge of Formosa in 1967.

Around the rest of the world, eels occur on most coasts which are washed by ocean currents that come from tropical latitudes. From such coasts adult eels migrate 'upstream' to warm spawning grounds and the currents drift the larvae back to the coast. Off the western coasts of the southern continents, off North America and off the eastern coast of South America, ocean currents originate from cold polar seas or are otherwise unsuitable for larval drift, thus no eels occur on these coasts.

The names, characters and distribution of the world's eel species of Anguilla as listed by Ege are given in Table 4. The main characters that distinguish the species are:

Colour: mottled or plain (7 species are mottled, 9 plain)
Number of vertebrae (averages range from 103 to 116)
Dorsal fin length (12 species have long fins, 4 species short fins)
Maximum sizes to which eels grow (range from 2.0 kg to 27 kg)
Habitat preferred (some like still-water ponds, some clear streams)
Distribution (see Table 4).

The range over which species are distributed varies greatly. The huge mottled [A.marmorata] (Fig 7) occurs over a huge 12,000 mile curve of this planet's surface from South Africa to the Marquesas islands in mid-Pacific (Fig 10). In contrast, [A.borneensis] is known only from eastern Borneo and Sulawesi and only five adults have been examined by scientists.

[Anguilla japonica] is the species of eel with the greatest number of vertebrae. It has 116 vertebrae on average. The common European eel [A.anguilla] has 115 vertebrae and the American eel [A.rostrata] averages 107 vertebrae. These three species of eels are very closely related and undoubtedly have descended from a common ancestor. Apart from the number of vertebrae and the pattern of their teeth, there is little difference between these three species.

GROWTH OF BROWN AND SILVER STAGE EELS
During their growing life in freshwater, eels are called

brown-stage eels (*yellow-stage* in American usage) on account of their brownish-yellow colour. Actually eels modify their colour to blend with their environment and clear rivers have pale coloured eels, dark peaty ditches have dark eels and truly yellow eels occur only in certain habitats.

The abundance of wild eels in different bodies of water varies greatly. It is determined by the amount of food available, competition and predation by other fish and by the numbers of elvers that reach that pond or stretch of river. Fast flowing rocky rivers have the lowest densities of eels, possibly only 5 kg per hectare of river bottom. Slow flowing muddy rivers, alkaline lakes, fen-land drainage ditches and brackish water coastal lagoons have the highest density of eels. In certain brackish coastal lagoons in Newfoundland, 400 kg eels/hectare (450 pounds eels/acre) are present.

Eels live from two to thirty years in freshwater before returning to the sea according to their species. Eels trapped in ponds or kept in aquariums have been known to live 85 years. The age of an eel can be determined by examining its otoliths (Fig 11).

In the last summer of their life in freshwater, eels change to what is called the *silver-stage,* preparatory to migrating back to the sea. Eels cease to feed as soon as they leave freshwater and thus must carry with them as oil a reserve of energy enough to power them on their long swim out to the ocean spawning grounds. Silver-stage European eels (the species with the longest sea migration) contain no less than 28 per cent oil by weight! The greater oil content of silver-stage than brown-stage eels causes silver-stage eels to fetch a high market price.

The main characteristics of eels in their two stages are:

Brown-stage *(yellow-stage):* body pigment dull, usually with a yellowy-green tinge, no glisten. Upper parts grey, brownish, greenish or yellowish. Underside dull white or grey. Fat content 5-15 per cent.

Silver-stage: A glistening layer exists underneath the skin. This causes the whole body to have a silvery glisten or glint, especially the white underside. Upper parts of the body are usually grey, occasionally with a purple sheen. No green or yellow tints remain.

In this stage the fat content is greater than in the brown stage and in the European eels reaches 25-28 per cent.

On autumn nights, especially during the last quarter of the moon, and when heavy rains have caused rivers to flood, the silver eels migrate downstream to the sea. By setting a net across a river, the entire run of migrating eels can be caught and, since adult eels never return from the ocean spawning grounds, the capture of these eels is totally harmless from a conservation point of view. Many other eels from lesser rivers that are not netted will always reach the spawning grounds and the elvers they produce are distributed by the ocean currents to the whole coastline (Figs 12, 13, 14).

FEMALE EELS ARE BIGGER THAN MALES
Female eels grow to much bigger sizes than do male eels and live longer in freshwater before returning to the sea. Since male eels are so small they have no commercial significance and 90 per cent of the weight of wild eel catches anywhere in the world is made up of females. The average sizes and ages of migrating silver eels of various species are given in Table 5.

FANTASTIC 6,500 KM MIGRATION
How European silver eels from many distant parts of Europe such as the Baltic, France and even from Egypt in Africa all migrate back to the single Sargasso spawning area is a mystery. They seem to swim near the surface, for the crews of lightships at sea sometimes see eels swimming past. The silver bellies of the eels are a camouflage protection against predators only when seen from below against the sky.

Johannes Schmidt noted that all eel species spawn in the highest salinity zones of the oceans. And eels are able to detect very small changes in salinity, light and gravity - especially the gravity of the moon.

After leaving river mouths and while still over shallow water, eels probably navigate using salinity and the light and the gravitational forces of moon and sun to guide them. Once beyond the continental shelf and into the ocean current system they probably just head into water of ever increasing salinity. This alone will automatically cause them to arrive at the spawning ground. Many must die on the way.

How many years does the journey take from Europe "up" the Gulf Stream to the Sargasso? The distance is about 6,500 km (4,000 miles) and a 7 km/day current is against the eels. Experiments on tagging silver eels in the Baltic show that they swim about 20 km/day. They could thus average 13 km/day against the Gulf Stream and it would take them 1 year 5 months to swim 6500 km. This time seems correct since it means that silver eels leaving European rivers in autumn would reach the Sargasso in the second February following - which is the month that spawning occurs. During this long journey they do not feed. They must be living skeletons when they arrive, and be hardly recognisable as eels. In most other species of eel the spawning areas are much nearer the coast, thus the migration back to the spawning area must be much quicker.

Although no-one has yet succeeded in catching eels at the ocean spawning grounds, it has been possible, by injecting hormones into female eels to induce their ovaries to develop almost up to full maturity (Fig 103). Many eggs can be seen in the enlarged ovaries of such eels and this has enabled estimates of fecundity to be made.

HOW EELS FEEL AND BREATHE

Along the flanks of an eel and on its head are rows of pores. This system of pores is called the lateral line system and each pore contains cells that are sensitive to vibrations. An eel has two tubular nostrils which open into large sensory pits. Elvers swimming close to a river bank are easily frightened away by heavy footfalls, for example. Since an eel moves and feeds mainly in the half-light of dawn and dusk, and often in murky water, it cannot see much with its eyes. Instead, the sensory cells of its nostrils and lateral line pores guide it, by enabling detection of smells and vibrations caused by any moving object.

Eels are smooth skinned and very slimy. Most people would say that an eel has no scales, but actually many very small scales are present but they are embedded under the skin.

Due to the absence of large scales, an eel can breathe through its skin as well as through its gills. The proportion of respiration (breathing) carried out through the gills is about 40 per cent and that through the skin about 60 per cent. On rainy autumn nights when silver-stage eels feel the urge to migrate to the sea, they can easily wriggle through sodden fields to escape from land-locked ponds into nearby rivers; as long as their skin is wet they can get enough oxygen to live.

The oxygen consumption of eels increases as the temperature increases, as is the case in all living things. This means that in winter a ton of eels have little requirement for oxygen and can be kept in a small pond. But in the heat of summer the same eels will need up to ten times the amount of oxygen.

MANY EEL MYSTERIES REMAIN

All the species of [Anguilla] undoubtedly have a similar life history, but the tropical species probably spawn quite near the coast and do not go to remote single spawning areas. The two Atlantic species are the "black sheep" of the family, living far cut off from the ancestral zone, shivering out their lives in cold waters, yet showing their true preference for tropical seas by making the most fantastic migration known in biology to reach the warm tropical waters of the Sargasso to spawn. How eels ever got into the Atlantic is itself a mystery. Probably [A.anguilla] and [A.rostrata] originated as wandering members of the Japanese eel species which somehow crossed from the Pacific into the Atlantic, perhaps through the sea channel which existed in the Panama region long ago.

Despite the solid foundation to ecological study of eels provided by Schmidt and Ege's work, there has been little progress in recent decades. The stage of knowledge at the present time is that the freshwater part of the life history - but not the marine part - is known in detail for the three northern eel species: [A.anguilla] and [A.rostrata] in the Atlantic and their close relative [A.japonica] in the N.W. Pacific. And right now, Dr. Peter Castle of Victoria University, Wellington, is leading an extensive investigation of the two New Zealand species.

But almost nothing has been discovered about the tropical eel species. And wild eels with eggs have still never been seen.

About the fascinating marine part of eels' lives hardly anything is known. Where are the spawning grounds of each species and at what depth? By what remarkable method do adult eels navigate from many different points to reach the spawning area - surely a topic with many useful repercussions to man. By what route and at what depth do they swim?

Do tropical eel species spawn all year round or have they a

single spawning season? Take the case of [Anguilla marmorata], the king of eels, growing to 27 kilos in size, whose realm extends from South Africa to Japan or to mid-pacific Tahiti and the Marquesas? (Figs 7 and 10). When and how does it spawn? Surely it must have at least three completely separate spawning areas?

Here are some projects that scientists might dream of tackling:

* To catch spawning adult eels of any species.

Probably it would be easiest to try one of the three northern species first, since we have a good idea as to the spawning place and season of these three. Delicate sonar, underwater lights, electric paralysing methods, underwater TV or submarines, depth charge explosions and huge nets may be needed - and money!

Probably the first man to see such eels will not recognise them as eels, they will be so emaciated.

* To discover the route and depth by which any adult eel travels from the coast to the spawning ground.

Here there should be studied a big species, to which could be attached radio transmitter tags; one that spawns near to the coast, so that the tracking ship would not have to be chartered too long at too great expense. A big [A.marmorata] of S. Africa or Java would be a good choice.

* To search the bottom of the Sargasso Sea to see if it is covered with countless bones of the eels that are supposed to die after spawning there, grabs and dredge nets would be needed.

Why do eels breed in the sea and enter rivers to feed whereas salmon breed in rivers and go to feed in the sea? Or is it connected with the fact that the density of animal life in the sea is greater in cold northern waters than in the tropics? Eels, being primarily tropical fish, may have discovered that there was little food available in the tropical sea, but plenty of food if they entered rivers. And maybe salmon found that highland rivers yielded a meagre diet but that the sea beyond the rivermouth was teeming with food.

There are many, many mysteries about eels that have yet to be solved.

3. The market demand for eels in Europe

About 22,000 tons of eels are eaten each year in Europe. Eels must be marketed alive or quick-frozen and glazed in order to secure top prices. Generally speaking, eels caught in Europe are sold alive, eels imported from U.S.A. and Canada are alive and eels imported from Australia and New Zealand are frozen. Unfrozen dead eels find almost no market. The price of eels is high and is rising steadily.

Who are the eel connoisseurs of Europe? In order of national consumption they are the Germans, Dutch, Danes and Swedes. In all these countries smoked eels are on sale in every fish shop and are eaten in every household, ranking as an expensive treat similar in status to sole or turbot in Britain. Actually the Dutch and Danes eat the most eels per person but the much larger population of Germany makes it the nation that eats the greatest quantity of eels. In Belgium, Italy (at Christmas) and in London, Britain, lesser amounts of eels are eaten. The eel consumption in Britain is at present less than about one thousand tons a year and is almost completely restricted to the east end of London. There is however a trend towards luxury eating in high-class clubs and restaurants with the price equalling that of smoked salmon.

In Europe all eels over 50 gm will find a sale but each country has its own size and type preferences (Table 3). In eating habits the Dutch are "suckers" but the Germans are "biters". When a Dutchman eats an eel he likes to feel the oil trickling out of the corners of his mouth and down his chin. When a German eats an eel he likes to bite something big and solid. In this way, Joh Kuijten the founder of the famous eel firm Joh Kuijten of Spaarndam, Holland, explained to his

29

sons the preferences of different countries. Holland and Germany are neighbours, yet strange but true, they prefer quite different types of eels for smoking.

Which are the main centres of eel buying and selling in Europe? In order of importance: Spaarndam, Holland (headquarters of the giant eel businesses of Kuijten and Kok), Hamburg and Copenhagen.

What factors affect the price of eels? The highest prices are obtained during the winter season December-April. This is when wild eels cannot be caught - thus supplies are low; the lowest prices occur during the main catching season, June-October. Silver-stage eels always fetch a better price than brown-stage eels of the same size, on account of the higher oil content and better flesh quality of the silver eels.

4. Life history of the Japanese eel *Anguilla japonica*

The Japanese eel [Anguilla japonica] spawns in the ocean near Okinawa. The larvae are carried toward the coasts by the sea currents and enter rivers as elvers; they ascend into fresh water rivers and lakes and live and feed there for 5-10 years until they reach adult size. After that they swim downstream into the sea and out to the spawning ground where they spawn and die. As in all species of [Anguilla] eel, adult females are much heavier than adult males.

The land distribution of [Anguilla japonica] extends from northern Honshu through southern Japan, Korea, Taiwan and down the coast of China as far south as Hainan.

The part of the life cycle that occurs in freshwater is well-known. Each year during December-April, millions of elvers enter river mouths from the sea. The elvers are about 6 cm long and are transparent (hence they are sometimes called "glass eels"). They prefer to enter rivers when the river water temperature is 8-10°C, thus the peak month of the immigration period varies somewhat between warm and cold years. The elvers have a strong instinct to swim against any water current and this causes them to swim vigorously upstream. When they encounter a waterfall they wriggle up the side of the falls among the wet mosses. They swim actively at night but in daytime hide under banks and stones etc. After some two weeks in freshwater, elvers become black in colour and look like small eels for the first time.

By midsummer, the young eels have reached rivers, ponds and streams all over the countryside and in the warm summer temperatures of 25-32°C feed actively on insects and worms and grow to lengths of around 15 cm.

31

This feeding and growth continues, in wild eels, for 5-10 years. Cultured eels grow much faster. Brown-stage eels is the name given to eels during this freshwater growing phase of their lives due to the fact that during this period their colour is brownish-yellow.

When over 30 cm in length, the sexual organs of eels can be distinguished for the first time. Finally, the eels reach adult size: males average 70 grams weight and 35 cm long, females are 300-350 grams weight and 57-60 cm long. Female eels are thus about four times the weight of males, and the ratios of males to females in quantity (numbers) are 4:6 in case of wild eels and 8 or 9:1 in case of cultured eels. (See Chapter 10 for reference to the feeding of hormones to eels to increase the ratio of females to males.)

In the summer of the year in which an eel reaches adult size nature prepares it for its journey down to the sea and out to the distant ocean spawning grounds. Great amounts of oil are stored in the muscles rising to about 20 per cent of the total body weight.

A metallic silvery colour develops under the skin of the eel giving the belly a silvery white appearance; the eels stop feeding, and on dark autumn nights the eels migrate down the river into the sea. Eels in this condition are called silver-stage eels.

5. Still-water and running-water — Culture methods

The principles of eel culture are the same as for all fish culture. Water, Elvers, Feed, Disease and Marketing are the five main aspects:

Water - for the eels to live in, a pure, plentiful supply is needed. Oxygen supply must be good and capable of being increased above the natural level for many eels to be cultured in a small area.
Elvers - The harvest of eels is nothing more than elvers grown to market size.
Feed - To grow the eels. The growth rate is determined by the temperature of the water and the amount of food eaten.
Diseases - and parasites must be prevented and/or cured.
Marketing - A man who knows how to market his eels will get twice the price for his fish compared with one who has no marketing experience.

Eels are cultured in ponds and live at much higher densities than can be supported by the oxygen supply available in an ordinary lake. To provide the extra oxygen that the eels need two fundamentally different methods exist and these give rise to two basic types of pond and culture techniques - the still-water and the running-water methods.

All Japanese eel ponds are of the still-water type and this is the method that can most easily be set up in most parts of the world. The running-water method (as used at trout farms) is not used at present for eels but may be used in the future, see Chapter 16.

Before going further we must clearly describe and

appreciate the characteristics of the two methods.

Still-water ponds: The first method of providing oxygen to the eels is by encouraging green phytoplanktonic algae to grow in the pond water which, by their photosynthetic activities, produce oxygen. Dense phytoplankton can only develop if the through-flow of water in the ponds is nil or very slow (if fast, the phytoplankton will simply be washed away). Thus ponds with more or less static water are required by this method. The daily through-flow of water in typical still-water ponds is about 5 per cent of pond volume per day.

Running-water ponds: The second method of providing oxygen to the eels is to let continuous new supplies of oxygenated water enter the ponds, as at a trout farm. Thus these ponds must be supplied with plentiful running-water all the time.

At any particular location, it is the volume and temperature of the water available that chiefly determines which type of pond is best:

When only limited water is available:
    still-water ponds
    (or recirculating running-water ponds)
Unlimited supply of water available:
    still-water ponds or in tropics only,
    running-water ponds.

TEMPERATURES OF 23-30°C NEEDED FOR COMMERCIAL SUCCESS

For commercial success, eels must reach market size of 150-200 gm in two years or less. Really fast growth is needed to achieve this growth rate and eels only grow fast at temperatures of 23-30°C. The time taken for eels to reach market size is one and a half years in Taiwan but four years in England (Table 18). Below approx. 12°C eels [A.japonica], [A.anguilla] and [A.rostrata] do not feed and thus do not grow at all.

Note that eels require much warmer temperatures for growth than trout: the water temperature suitable for eel culture is 23-30°C, and for trout 10-15°C.

Thus out-of-door unheated ponds can only be used for commercial eel culture in tropical and sub-tropical areas such as, Indonesia, Taiwan, southern Japan, Madagascar, the

Caribbean, Queensland or Tunisia.

In cooler areas such as northern Europe, southern Australia and New Zealand, eel ponds must be heated artificially if eels are to be grown commercially. There are several ways of doing this:

* Still-water ponds; covering the ponds with greenhouse covers of polyethylene or glass (Fig 115).
* Still-water ponds; Laying pipes on the pond bottom through which hot water is circulated to warm the pond, i.e. pond central heating.
* Running-water ponds; Using covered ponds with insulated bottoms and walls in which the water is heated, purified and recirculated. This will be very expensive but is being attempted experimentally.

6. Method and organization of eel culture industry in Japan

Two species of eel are cultured in Japan; the local species [A.japonica] (95 per cent of the cultured harvest in 1971) and the European species [A.anguilla] imported as elvers by air freight (5 per cent of cultured harvest in 1970). The import of European elvers to Japan started about 1969 and they grow equally as well as the Japanese species except that they do not like the 30°C temperatures which occur during the height of Japanese summer.

The eel ponds of Japan are not spread evenly around the country but are nearly all concentrated in two areas (Figs 15 and 16). The biggest is around the shores of the brackish lake Hamanako near Hamamatsu in Shizuoka Prefecture (*Hamana* = name, ko = lake, *Hamanako* = 'Lake *Hamana*'). The other is Yoshida district by the Ohi river in Shizuoka Prefecture about 100 miles south-west of Tokyo, extending to Yaizu district. Both these areas are beside the coast.

All the eel ponds in Japan obtain their water supply from shallow bore-holes and use the *still-water pond system.* (Fig 37.) The reason the still-water system is used is as follows:
Bore-hole water is 15-20°C all the year round which is cold for eel culture. However, in still-water ponds the sun heats the water up to 25-30°C and this is ideal for eel culture. This still-water enables phytoplankton to propagate which gives suitable water darkness and increases oxygen in water thus providing suitable conditions for eel culture.

In certain isolated eel farms in Japan, and in most eel farms in Taiwan, river water is used to irrigate the ponds. These farms also use the still-water system. Young elvers, however, are reared in running-water tanks.

Eels are raised from elvers to market size using four sizes of ponds (see Table 6). As the eels grow they are known by different names which are described in Figs 17 and 18. The methods used to feed the eels of each size are described in the chapter on feeding. Figs 18 and 19 show the weights of eels of each different length. The maximum density at which eels can live in the most modern ponds is 4 kg/m^2 (Table 7).

Elvers at the transparent "glass-eel" stage are the starting point of eel culture. As yet, no-one has succeeded in breeding eels and, as the natural spawning ground is far out to sea, it is necessary for elvers to be collected when they enter river mouths. They are about 6 cm long, transparent and weigh about 0.16 grams each (it may take up to 6,000 elvers to weigh one kilo). On arrival at the farm they are put for half an hour in a bath of Nitrofurances to disinfect them of any adhering bacteria.

The elvers are placed in the elver tank at a density of about 400 gms per m^2. 20-50 per cent mortality of elvers may occur during the first month. It is important that weak, stunted or diseased elvers be removed as soon as possible.

After 20-30 days in the first pond, the elvers are caught, sorted into groups of two sizes and transferred to the second elver ponds. All undersized and stunted elvers are separated. In the second pond stocking density is about 100 gm elvers/m^2. After a further 20-30 days the elvers, now about 12 cm long are caught again, sorted by size and transferred to the outdoor fingerling ponds.

In June-July, the fingerlings are caught and put in adult ponds. Many fingerlings are sold to other farmers at this stage. In August-September when 20-30 cm long, the eels are caught, graded by size and the large and small sized eels separated into different ponds, in which they over-winter and live until they reach market size at the end of the second summer growth. The rate of weight increase in each month is shown in Table 8. A typical eel farm is shown in Fig 20.

How long does it take for cultured eels to reach market size is a question often asked. It is difficult to answer this question because the rate of growth varies between ponds and from farmer to farmer according to the efficiency of his operations. And each year eels grow faster as better techniques are introduced. At present most eels reach 60 gms by the end of the first growing season and at the end of the

38

second growing season reach marketable-size of 150-200 gms. But even now some farmers in southern Japan harvest 120 gm eels in November of the same year in which the elvers were caught. Many eels are silver-stage when they are harvested.

In the second growing season, the young adult eels quit hibernation when the temperature rises above 12°C in April. They are fed once per day in the morning, reach market size in the autumn or early winter and are marketed during the winter months, October to March.

The annual weight budget of eels in a typical 200 m² adult eel pond is approximately:

Weight gain:
Under sized eels carried over from previous year	200 kg	
Fingerling eels stocked in August	100 kg	
Meat weight gain due to feeding	700 kg	
		1,000 kg

Weight loss:
Winter harvest of eels	800 kg	
Undersized eels left to grow	200 kg	
		1,000 kg

General rules that will promote good results in eel culture are:

* Keep your eels in the best possible health. Thus careful control of water quality, avoidance of parasites and disease, and avoiding damaging the skin of eels when netting or transferring them will all help.
* Continually sort the eels during their growth, dividing the different sizes so that in any pond the eels are of almost the same size. Separate all undersize or stunted eels. This is one of the main techniques of successful eel-farming.

Given normal mortality, 1 kg of elvers will yield a harvest of 400 kg of market-size eels, i.e. if you desire a harvest of 40 metric tons of eels, you must start off with 100 kg of elvers.

The Japanese have tried out many variations of culture methods, pond designs and feeds during the past 80 years. Rather than describe the successful methods in one chapter, this book now presents the techniques in a step-by-step manner so as to guide persons who live in Europe, Australia and New Zealand or any other suitable locality on how to start eel farming from nothing.

The success of eel culture in Japan is based on sound business management as well as on good biological techniques. The main divisions of the eel culture industry and their relationships are shown in Fig 21. When eel culture first started, the pioneers caught their own elvers and raised them through to market size. Now the position is much changed.

Elvers are caught at river mouths all around the southern Pacific coasts of Japan but supplies can no longer meet the demand and additional elver supplies are imported by air freight from Europe.

Many farmers grow eels only from 20-30 cm fingerling size up to market size. Thus other smaller farms have developed which rear elvers to fingerling size, and then sell their fingerlings to the other farmers.

In each district, eel farmers' co-operatives have long been in existence and all farmers are members. The co-operatives do many useful things such as:

* own cold stores
* buy artificial feed and fish in bulk at wholesale prices for the farmer members.
* market the harvests of farmers, then pay each farmer his share of the total proceeds.
* negotiate with the Government etc.
* assist members with accounting and legal matters.

Additional to all the private enterprise organisations are well-equipped government research and experimental stations in each area. And at many technical colleges and several universities, students can take training courses in aquaculture.

The eel farming area, about 100 miles north-east of Tokyo, is within the Pacific Ocean zone. There are no large eel-culture ponds in the estuary of the Tone river. There are very few eel-culture ponds in the neighbourhood of Tokyo, but in the areas south of Tokyo there are many more eel-culture ponds than in the north. There is only one large eel-culture pond about 40 miles (65 km) East of Tokyo: namely Hikari machi in Chiba Prefecture along the Pacific coast. This pond is 10.4 ha. in area and produces 73 tons of eels annually.

Concerning pond areas, the difference between Hikari machi and Shizuoka Pref. is shown in figs 15 and 16 and the following table gives the latest details of units and areas in the Shizuoka Prefecture.

Management unit and area in Shizuoka Prefecture
(as of April 1973)

district	management unit number	water area
Shida	106	148 ha.
Haibara (Yoshida)	403	223 ha.
Chuen	34	60 ha.
Seien	320	786 ha.
Total	863	1,218 ha.

The leading position of the Shizuoka prefecture in production of cultured eels is shown in this table giving the tonnage of eels from the areas farmed in the various prefectures during 1972:-

Prefecture	Production (tons)	Area $(1,000m^2)$
Shizuoka	7,919	13,299
Aichi	3,950	6,921
Mie	787	1,742
Miyazaki	262	874
Tokushima	261	1,768
Chiba	222	696
Kagoshima	205	544
Okayama	157	365
Fukuoka	91	159
Shimane	53	115
Kumamoto	43	148
All Japan	14,460	26,629

7. Choosing a site for an eel farm

A good site for an eel farm must satisfy as many of the following conditions as possible:

* a good supply of water, channelled or pumped from a river or pumped from boreholes. About 450 m^3 per day of water are needed to grow 20 tons of eels per year.
* water may be clear or turbid, but must not be liable to poisonous pollution by insecticides etc. Alkaline or neutral water is best. Acid waters are not suitable for eel culture. Plenty of wild eels living in the water you propose to use is a good sign - it proves its suitable quality.
* not subject to flooding.
* high enough above adjacent outlet level so that ponds can be emptied by gravity alone.
* non-porous soil, so that water does not leak out of ponds; sandy clay is best.
* sunny position to encourage growth of oxygen-producing phytoplankton.
* open windy position to encourage oxygenation of the surface water by the wind.
* roads and electricity available.

In Japan, certain flat lands adjacent to rivers and brackish lagoons have been developed into huge eel culture centres. For reasons of both psychology and economy, Japanese eel farmers prefer to locate their farms adjacent to existing farms and this is certainly beneficial. Since the eel farms are so concentrated, it has been easy for co-operatives to develop. (They offer cut price central frozen-feed stores, equipment supply and accounting, subsidiary and marketing services.) This concentration also makes it convenient for the government to set up experimental stations in each area and to offer free disease-prevention services etc. to all farmers.

In Europe, flat lands adjacent to big rivers or estuaries would be suitable to develop into eel farms or, for single farms, sites near disused water mills offer the advantage of gravity-fed no-cost water supply.

8. Eel farm design and construction

All Japanese ponds have the "still-water" type of irrigation. Running-water ponds are discussed in Chapter 16. Figs 22 to 29 show many details of typical Japanese eel ponds. The trend nowadays is for ponds to become smaller and smaller. At present (1973) cultured eels are accommodated in a sequence of four pond sizes as they grow from elver to market size (Table 6).
1. First elver tank, usually under a greenhouse.
2. Second elver tank, usually under a greenhouse.
3. Fingerling pond, outdoors.
4. Adult eel pond, outdoors.

ELVER TANKS
Elver tanks are usually built under greenhouses and are heated to above 25°C by thermostatically-controlled electric immersion heaters. This heating gives the elvers a fast start in the spring; a technique also used in trout and oyster farming. The first elver tank into which the wild-caught elvers are placed and reared for a month is usually circular, made of concrete 5 m diameter by 60 cm deep. Water is sprayed onto the surface from angled jets creating a circular whirlpool current. The water drains away through a central pipe. The second elver tanks raise the elvers from 8 cm up to 12 cm size and are about 30-100 m² area and 1 m deep. Both types of elver tanks are aerated by bubblers as well as by the surface jets of incoming water.

It is a good idea to obtain the water supply for elver tanks from a bore-hole or well. This guarantees a pure supply for the delicate elvers. Larger sizes of eels are more robust and can tolerate the variations of water quality that may occur sometimes in river water.

FINGERLING PONDS

Fingerling ponds are usually rectangular, of about 200-300 m^2 in area, and 1 m deep with a mud bottom. To prevent the escape of small eels on rainy nights when rain trickles into the outdoor pond it is desirable for the concrete pond edge to have an overhanging lip for elvers and fingerling eels. Above the length of 20 cm eels do not try to leave the pond, thus in adult eel ponds no overhanging lip is necessary.

Adult eel ponds are rectangular or square and were formerly made between 5,000-20,000 m^2 (0.5-2.0 hectares) in area, as shown in the series of Figs 22-29. Recently the size has become greatly reduced to only about 500-1,000 m^2 in area and may be further reduced in future. Increased sophistication of management is needed in the care of small ponds.

By the running water culture method and the use of efficient drainage with the water at temperatures of about 28°C, marketable-sized eels to a total weight of 1.5-2.5 tons (10 kg/m^2) can be produced in a concrete pond (with a concrete bottom) of 1 m in depth and of 15-18 m in diameter.

In Japan and Taiwan at the present time, adult eels are raised in several types of ponds, dating from different periods of construction. Large ponds, in some cases completely stagnant, are used in some areas where the water supply is small but where flat land is plentiful and cheap.

A drainage ditch about 1 m wide and with its bottom at a lower level than the pond bottom, is excavated adjacent to the exit sluice of each pond to carry away discharged water.

CONSTRUCTION

Ponds are usually constructed by excavating the bottom of the pond using a bulldozer, piling the excavated earth around the pond to form banks, and then facing the inward side of the banks with concrete slats or cemented stones to prevent them from being eroded by waves. Banks between ponds are usually about 2 m wide at pond-bottom level and 1.4 m wide on top. Table 10 shows the costs of constructing two large ponds in Japan in 1967.

Each pond has one main water inlet and one main water

outlet. The inlet is either a pipe bringing in pumped water or a sluice leading water in from a river. If the incoming water can be poured into the pond from a height of 60-100 cm extra oxygenation will occur. Such pouring from a height can easily be arranged when piped water is used.

The bottom of the pond is built to slope slightly toward the outlet sluice and is shallowest - about 80 cm deep - near the water inlet and deepest - about 120 cm - near the water outlet. This allows the pond to be completely drained by simply opening the outlet sluice.

Various photographs of ponds during and after construction are given in Figs 30 to 35. Sluice gates at the outlet of each pond consist of a concrete lined gap in the pond wall bearing slots for three different kinds of doors (Fig 33). To cater for the various operations of pond management the three types of doors are vital and they fulfil the following functions:

1st slot: board door: this permits the pond operator to choose whether surface or bottom water layers of the pond are discharged.

2nd slot: mesh door: this prevents the escape of eels.

3rd slot: sluice door: this controls the amount of water leaving the pond.

The size, strength etc. of a given inlet or outlet sluice gate will vary according to the volume of water flow with which they have to cope.

On the outside of pond outlet sluices the water is discharged through a big pipe into the drainage ditch. By tying a bag net onto the end of this pipe, eels can be harvested when the pond is being drained (Fig 35).

An alternative method of permitting water to escape from ponds is by the use of 20 cm diameter plastic tube siphons set across the pond bank (Figs 36, 45 and 46). A net cylinder around the inlet end prevents fish escaping; a hand pump applied to an air valve at the top is used to exhaust air from the pipe and set the water siphoning over. The outlet end of the pipe discharges into the drainage ditch.

Electric wires supplying the pumps etc. should be buried in the pond banks if possible, or otherwise hung from poles. It is desirable that electric switches controlling pumps and aerators etc. should be installed all together at the main watchman's hut.

To prevent the banks from collapsing, they must be faced in some durable way. The most common method is to use slats and grooved posts of concrete (Fig 25). The grooved posts are set at 1.3 m intervals and the slats, which are exactly 1.3 metres long, 3 cm thick and 35 cm wide, are slotted into the grooves to make a solid wall. Boulders cemented to form a wall are also used but this requires more labour and expense. The facing of the banks should extend from 30 cm below the level of the pond bottom to 50 cm above the pond surface.

Ponds for adults and fingerlings have a special feeding place incorporated in one bank of the pond (Figs 38 to 41). The feeding point should be sited on the side of the pond opposite to the source of the prevailing wind. Such waves increase oxygenation of the water which in turn increases the eels' appetites. The feeding point consists of a built-out "bulge" from the main bank, which presumably deflects the water flow of the pond and thus makes it easy for eels to locate it. In some cases a shed or platform of some sort is erected so as to shade the feeding point for the reason that eels like darkness.

Many farmers construct a "resting corner" in the corner of adult eel ponds that are larger than 600 m² (Figs 42 to 44). The resting corner is really a running-water pond linked to the main still-water pond. The resting corner has a water inlet and outlet, a splasher-aerator machine and a vertical pump, and has a water circulation that is almost separate from the water in the main part of the pond and is of particularly high oxygen level and purity. It connects with the main body of the pond by a small opening only. Thus, if water quality is bad in the main body of the pond, such as at night when the oxygen level falls, the eels enter the resting corner and find better conditions. Likewise, when eels are netted during harvesting, the dense bag of eels is dragged to the resting corner for sorting etc., and the higher oxygen supply in the corner saves them from dying of lack of oxygen. A small opening in the wall dividing the resting corner of the pond from the main part of the pond allows eels to enter or leave as they wish.

A typical resting corner occupies two per cent of the area

of these large ponds. At night, extra water can be pumped into the corner, the aerators switched on and oxygen thus kept high until morning.

9. Water quality and its control

As air is to humans, so is pond water to the eel. If the quality of the pond water is good, eels will be healthy and grow fast. If it is poor, they will be unhealthy and grow poorly.

The key to good water quality is *good circulation,* therefore the shape of ponds and the positioning of inlets, outlet sluices and paddle machines must ensure that all the water of the pond circulates freely. If any stagnant water exists it will be sure to cause trouble.

The characteristics of the water that are most important are the oxygen content, pH, nitrate, temperature, and the amount of phytoplankton. (For details of fast-flow types of ponds which have no phytoplankton see Chapter 5.) The water in a pond is subject to many influences and none of these characters stay static but vary during each cycle of 24 hours, from week to week and from season to season. The feeding and breathing of the eels, the growth of plankton, the running in of new water and running out of old water, all influence the water quality. But most of all it is the growth of the phytoplankton in relation to the rising and setting of the sun each day which affects water quality.

The normal daily variation of oxygen, pH and temperature in a pool is shown in Fig 47. Oxygen ranges between 1 and 10 ml/litre and pH between 7 and 9. Nitrate in healthy ponds varies from 0-100 mg/litre during the course of each phytoplanktonic bloom and decline.

Typical eel ponds are a thick soupy green colour in summer and this growth of phytoplankton is encouraged because by day the life processes of the phytoplankton produce much oxygen which augments the natural supply in the water and allows large weights of eels to live in the pond. At night, however, the phytoplankton use up oxygen and on hot summer nights the oxygen levels of ponds frequently drop below the level at which eels can breathe. Certain species of phytoplankton are dangerous, e.g [Anabaena] blooms reduce the appetites of eels greatly. Zooplankton are not desirable in eel ponds. Blooms of zooplankton can cause severe oxygen depletion. A healthy pond contains phytoplankton and zooplankton in proportions about 98:2 by weight.

Phytoplankton do not exist in a steady stage of abundance but go through cycles of growth, reproduction, spore production and death. These occur about every 40 days in summer and the colour of the pond and its water quality change according to the stage of the cycle (Fig 48). A *bloom* is the name given to the sudden appearance in a pond of dense masses of plankton. When a bloom or a sudden death of plankton occurs, the chemical balance of the pond is temporarily upset and there is danger of oxygen depletion. The transparency of a pond is a measure of the density of phytoplankton in the water and should be maintained between 10 and 40 cm if possible, as determined with a white disc lowered below the surface.

Oxygen is the most critical water characteristic. At oxygen levels below 1 ml/litre eels cannot breathe and rise gasping to the surface. A good eel farmer frequently checks his ponds, especially at night, to see if any fish are swimming gasping near the surface. If he sees any, it is a sure sign that urgent action is needed to increase the oxygen supply. It is a good idea to test the oxygen content of the bottom water of the pond also and use vertical pumps to raise it if necessary (Fig 49).

Oxygen deficiency regularly occurs under the following conditions:

* on summer nights between 8 pm to 6 am, during which time the phytoplankton use up much oxygen.
* When many eels are concentrated together, such as when harvested in a net.
* When deaths of phytoplankton occur, or when harmful species of phytoplankton multiply in the pond.

To improve the water quality in eel ponds there are several methods:
* Let more water than usual flow into the pond, thus increasing the rate of water exchange and letting foul water pass out of the pond.
* Switch on paddle-splasher machines. These splash water into the air, thus oxygenating it and improving horizontal circulation in the pond (Figs 50, 51, 52).
* Switch on vertical pumps. These draw water from the lower levels of the pond and shoot it out over the surface, thus oxygenising the water, improving vertical circulation in the pond and preventing accumulation of foul water on the pond bottom (Figs 53, 54).
* Add various chemicals to the water such as CaO to reduce acidity, iron oxide to remove Hydrogen Sulphide.
* In extreme cases a pond must be emptied, dried out, disinfected and refilled with new water.

A great range of models of splasher-paddle machines and vertical pumps exist in Japan and a selection of modern ones is shown in Figs 50 to 55.

The modern concept of eel culture is to grow as many eels in as small a volume of water as possible. This causes the chemical balance of the pond to be very precarious and demands sophisticated quality control. Oxygen deficiency arises regularly during summer and on many farms the paddle-splasher machines and vertical pumps are run continuously during warm weather.

Each morning and evening farmers test the oxygen, pH, nitrate, plankton etc. levels in each pond and switch on the aerator machines or adjust the rate of water inflow to bring about optimum water quality. Some simple equipment and a microscope are needed for doing the tests.

Most Japanese eel farmers sprinkle powdered lime and Ferric oxide (Fe_2O_3, trade name *Manken*) on the surface of ponds every 1-2 weeks. The former aids decomposition of organic waste, excretions etc. in the pond and the latter aids phytoplankton growth and destroys any harmful sulphides in the water by combining with them to form an insoluble precipitate of Ferrous Sulphide.

Rotifers, [Brachionus plicatilis], are one of the most dangerous zooplanktonic animals that occur in eel ponds.

These minute swimming animals feed on phytoplankton and use up much oxygen, thus wholesale asphyxiation of eels may occur. When microscopic study of a water sample reveals more than three rotifers in the field of view the only course of action is to completely change the water in the pond as fast as possible.

"Water flea", [Cladocera] plagues sometimes occur in ponds. These zooplankton can easily be killed by sprinkling *Masoten* on the water.

10. The feeding of eels

In Japanese farms artificial feed and raw fish are used to feed eels and the use of artificial feed is rapidly increasing. At present about 70 per cent of farmers use artificial feed and 30 per cent use raw fish. In Japan feed amounts to about 30 per cent of the production cost of cultured eels.

The conversion ratio of raw fish and artificial feed is shown in Table 11. The apparently great difference is due to the fact that artificial feed is dry. The conversion ratio of both types of feed is higher in warm temperatures than in cold, and among young eels than old eels (Figs 56, 57, 58, 59).

Artificial feed is convenient to use because it does not require cold storage and needs little space. It is produced in powder form and is mixed with water and 5-10 per cent of vitamin oil to make a paste (Figs 56, 61). This paste is placed on a mesh tray in the eel pond and the eels go through the mesh or climb on to the tray to get the food. Uneaten paste remains on the tray and is lifted out of the pond, thus does not foul the pond. Artificial feed is made chiefly of fish meal with added carbohydrate and consists of about 52 per cent protein, 24 per cent carbohydrate, 10 per cent water, 4 per cent fat and 10 per cent ash. Artificial feed caused enlargement of eel livers when first introduced but its composition is now amended to avoid this. The percentage of vitamin oil added to the feed is varied according to the water temperature: below 18°C about 5 per cent by weight is added, above 18°C about 10 per cent is added.

Raw fish used are mackerel [Scomber], launce [Ammodytes], pacific saury [Cololabis], ataka mackerel

[Pleurogrammus], horse mackerel [Trachurus], sardine [Sardinops] and anchovy [Engraulis]. These fish are usually bought in frozen blocks, then thawed overnight and fed whole to the eels. Offal is not used because it has poor nutritive content and tends to cause outbreaks of disease. The analysis of various fish is given in Table 12. For machines for food preparation see Figs 60, 61.

It is vital that no unnecessary organic debris be allowed to fall into ponds, thus when feeding eels, the food is not just thrown into the pond. Instead, a special feeding point is established at one side of each pond and the eels learn to come there each feeding time. Artificial feed is lowered on a perforated tray. Raw fish are threaded through their eyes on a piece of wire, dipped in a tureen of boiling water for a minute to soften the skin and then lowered into the pond. After the eels have eaten off the flesh (an orgy lasting less than five minutes!) the wire is raised, bringing out the head and backbone of each fish, and these are thrown away. The only organic matter that enters the pond is what the eels actually eat.

Eels are fed once only per day, about 8-10 am and in the warmest part of the summer the quantity of feed given should be about ten per cent of the body weight of all the eels in the pond. If more than this amount of food is given conversion efficiency decreases and money is wasted. In cool months feed is reduced and during winter no food at all is given.

The quantity of food consumed on a typical eel farm each month is shown in Table 13. In spring little food is used and then during the summer consumption rapidly increases for two separate reasons:

1. The warmer temperatures encourage eels to eat much.
2. The total weight of eels in the ponds steadily increases as they grow.

Naturally the appetite of eels in the ponds also steadily increases as they grow. They eat most on clear, windy, dry days and eat least on rainy, cloudy, calm days.

A few years ago, harvests were sometimes found to contain a preponderance of small eels and on examination these were found to consist mainly of male eels, which grow to much smaller sizes than females. Eels are unusual in that their sex is not finally fixed until they are over 30 cm long and the above phenomenon is thought to be caused by the crowded conditions of culture somehow influencing more than the normal percentage of eels to become male. The remedy is to mix some hormone with the food of eals, which tends to make them all grow to be females.

Elvers do not take food for the first few days after capture and will only gradually develop an interest in feeding. First feeding of newly caught elvers is done just after dark, using a bright electric bulb suspended over the water to attract the elvers to the feeding point. Care must be taken not to put in the pond more food than the elvers will eat or it will decay. The usual method is to put the feed paste on a gauze tray and lower the tray just beneath the surface at the edge of the elver tank. Let the elvers eat their fill, then remove the tray and any uneaten feed still in it. Below 13°C, elvers do not feed, thus elver ponds must, artificially or naturally, have temperatures above this value.

Elvers are fed a sequence of diets as below:
Weeks 1 and 2: minced bivalve flesh or minced earthworm flesh
Weeks 3 and 4: minced fish flesh/Tubifex or similar worms
Weeks 5 to 10: minced fish flesh or artificial feed
New artificial feeds are being developed acceptable to elvers.

The elvers are fed all day, from 8 am to 3 pm and eat about 25 per cent of their body weight of food per day.

It is not possible to give elvers or large eels one type of food one day and another type the next day; all changes of diet must be introduced gradually by mixing in increasing proportions of the new type of food over a period of about a week.

KEEP AN AQUARIUM
Keep some elvers in an aquarium. This is good advice to all eel farmers. Because in an aquarium you can actually watch what the elvers do, what they like and dislike and how they behave. After raising elvers in an aquarium you have much useful knowledge to guide how you treat the larger numbers of elvers you wish to culture in tanks - where you cannot observe them closely.

The phases of behaviour of elvers are roughly as follows:
Newly caught up to the 4th day: The elvers hide in crevices,

under stones etc., emerge only at night (when they are attracted by a light), and show no interest in food.

4th-10th day: The elvers hide in crevices, become grey in colour and begin to feed. The best food is mashed earthworms. Dig fresh garden worms, on a wooden block cut them fine with a razor blade, then grind into a paste using the back of a spoon. At first when mashed worm meat is dropped in the aquarium, no elvers take any notice of it unless they actually bump into it by accident. When an elver does this it starts to make feeble bites at the bottom, biting surrounding stones and finally by accident bites the meat fragment. It takes the elvers an hour or more to eat their feed, given once daily.

10th-20th day: The elvers swim actively for about half their time, hide in crevices, among weeds or bury themselves in the gravel with just their heads sticking out the other half of the time. They are darker in colour but their bellies are more or less transparent. After feeding, it is easy to see the food in their stomachs. They feed twice daily at 0900 and 1700 hours and the food is always placed in the same corner of the aquarium. Elvers now quickly detect the presence of food and are excited by the smell, but are slow to locate exactly where the food is. As the current diffuses the scent of the food around the aquarium, the elvers suddenly start to swim vigorously or prowl searching along the bottom. They locate the food entirely by smell (blind eels kept in aquariums have lived many years) and grab it greedily when they find it.

21st-30th day: The elvers are now noticeably growing and it can be seen that some grow faster than others. They are now used to their feeding routine and as soon as the food is put in the aquarium, they quickly smell it and come to the place and greedily grab the food. Each feeding session is over in ten minutes.

When your elvers get this far, you are well on the way to success. They become ever greedier and the four vital aims one must have are: 1. Keep the water crystal clear.

2. Give the elvers as much food as they can take; they therefore grow quickly.

3. But never let surplus uneaten food get into the tank or it will poison the water and cause disease. Thus food is not thrown into the pool but is offered on a tray suspended just under the surface. When elvers have had enough, any

surplus is lifted out of the tank.

4. Throw out stunted or undersize elvers whenever they are detected and periodically separate the remainder into fast and slow growers and raise these two groups separately.

Good feeding practice in the early feeding period up to six weeks of age is of great importance as upon the progress in this period depends the final success of the operation. This is in line with the recognised feeding pattern of young life - progress in the early stages lays solid foundation for steady progress later. In one experiment the writer (Mr. Usui) achieved in four months a weight of 10,000 grammes for one batch of elvers compared with five months taken for a comparable batch to achieve a somewhat similar weight by a different feeding programme.

11. Diseases and parasites

Eels are attacked by a considerable range of diseases and parasites and the main ones are listed in Table 14 and illustrated in Figs 62-70. Many of the diseases are temperature-dependent, i.e. they only flourish within a certain range of water temperature. Apart from the listed attackers, eels can suffer from malnutrition due to an unbalanced diet.

Prevention is better than cure so the best advice to eel farmers is 'ensure diseases never get a chance to strike in the first place'. Table 15 lists medicines used by Japanese eel farmers.

The main methods of preventing a disease and parasites of eels are:
* ensure ponds always have a pure and plentiful water supply.
* dry and sterilize ponds after each harvesting. On dried out and on full ponds scatter lime and ferric oxide to decompose organic wastes on the bottom.
* give good, balanced diet, sometimes with added anti-bacterial drug etc.
* avoid damaging the skins of eels when they are handled.
* during the last week of the autumn in which feed is given and during the first week of feeding in spring, add anti-bacterial drugs and vitamin E to the feed.
* kill intermediate hosts of parasites by adding suitable poison to the pond water.
* remove dead eels quickly from the pond and burn the bodies. To cure diseased eels is difficult. It is not practicable to catch eels and transfer them into a medicinal

dip. The best treatment when eels are found to be diseased or parasitized is as follows:
* add medicine to the food of the eels.
* add medicine to the pond water.
* pump salt water into the pond (in certain cases).

Knowledge of most diseases is rudimentary at present and in the case of some diseases, all a farmer can do is hope for luck, isolate any ponds in which disease occurs, remove eels that die and hope that some survive. In serious epidemics all the eels in a pond may die or may have to be killed, the pond dried and sterilised and a fresh start made with new fingerlings.

FUNGUS DISEASE

This is the most serious disease because no effective cure has yet been found (Figs 62, 63, 64). In 1966 and 1968 this disease struck the main eel pond area of Japan (around Hamanako to Yaizu district) and several hundred tons of eels died.

The disease breaks out during periods when the water temperature is 15-20°C and thus in Japan two peaks of infection occur, one in spring, one in autumn. Patches of white fungal growth develop and spread on the bodies of eels and cause death after one or two weeks. 50-70 per cent of eels in a pond may die.

To prevent outbreaks of white fungus disease the following measures may be used:
* in elver ponds sprinkle in some Nitrofuran compounds.
* add prophylactic anti-bacterial drugs to the feed.
To cure saprolegniasis the following are used, but most of the diseased eels cannot be cured:
* for one week add anti-bacterial drugs to the feed, e.g. Daimeton at 150 g/ton of body weight of fish.
* put Malachite green into the pond water, four times at 10-days intervals (0.2 ppm).
* put Methylene blue into pond water for three days (2 ppm).

RED DISEASE (Fig 68) is a bacterial disease mainly affecting large eels and occurs when water temperatures are 18°C or higher in the summer. The disease causes the fins and body to become raw and red and, internally, the intestines, liver and kidneys are affected. The disease sometimes originates from the use of decaying food, so care must be taken to keep the diet fresh. To cure infected eels:

* add anti-bacterial drugs to the feed for five to seven successive days (give 100-200 g of medicine per 1 ton of eels).

SWOLLEN INTESTINE DISEASE is a bacterial disease in which the intestines become swollen.

GILL DISEASE (Fig 69) is caused by bacteria which attack the gills and cause them to rot and death follows. The addition of anti-bacterial drugs to the feed is said to cure gill rot in some cases.

BRANCHIONEPHRITIS causes gills to be swollen, raw, rotten etc. and kidneys to be inflamed. The cause of this disease is unknown at present. The cure for this disease is also not settled but it is good to keep the pond water to 0.5-0.7% in salinity. During 1969 and 1970, this disease struck the main eel pond area (about 90%) of Japan, killing 2,600 tons of eels.

WHITE SPOT DISEASE (Fig 67) is caused by [Ichthyophthirius multifiliis]. The cure for white spot disease is to pump salt water into the eel pond and let the eels live in salt water for a few days.

MYXIDIUM-DISEASE is caused by a protozoan parasite called [Myxidium]. Spores settle on the skin, develop into a round white expanse and grow larger. The course to take for this disease is to remove the diseased eels and burn them.

CRIPPLE BODY DISEASE (Fig 65) is caused by the protozoan parasite [Plistophora] which attacks the muscular system causing the victim to become crippled and misshapen. In small eels, the diseased part of the muscles can be seen through the skin as a milky colour.

GAS DISEASE: this disease attacks eels, causing bubbles on the head (Fig 66) and occasionally on the fins of adults. The condition is liable to occur if the O_2 or N_2 become excessive. The bore-hole water, which brings an excess of N_2 in the pond should be checked previously before use. The cure for this disease is to run new water in the pond or to work the water agitator for reducing O_2.

ANCHOR WORM DISEASE (Fig 70) is a common and important parasitic infection caused by the copepod [Lernaea cyprinacea]. These animals spend their lives attached to the inside of the mouth of the eels (where they greatly hinder the eel's ability to feed) and attached onto other parts of the outside of eels' bodies. The heads of the parasites are embedded in the flesh of the eel, sucking out the juices. Heavy infections cause death due to preventing the eels from feeding.

Adult [Lernaea] release up to 5,000 eggs each, which hatch into free-swimming larvae and these larvae swim about until they meet an eel when they hang on and the story is repeated. During one summer the life cycle is repeated about five times. The most suitable propagation condition is with a water temperature of 14-32°C; below 14°C the parasites do not reproduce. Since [Lernaea] also live on carp etc. care should be taken that any carp kept in eel ponds are disinfected before being put in.

To cure anchor worm disease:
* pump sea water into the ponds and let the eels live in sea water for 3-4 days. This kills eggs and larvae of the parasite.
* Sprinkle calcium chloride powder over the surface of infected ponds just after dawn. The eggs and larvae float near the surface and are killed by the chlorine which disperses and does not harm eels at the bottom.
Masoten added to ponds at concentration of 0.2-0.5 ppm kills larvae of [Lernaea,] which float on the water surface at dawn owing to their phototaxis. For prophylaxis, Masoten can be applied routinely twice a month during the summer months.

In addition to the above diseases, various internal and external parasitic nematodes and trematodes also attack eels, but are not important.

In Europe, live eels stored in perforated barges in fresh water at Maldon, Essex, in summer get a disease called 'red sickness' in which the fins become tinged red, the body hard and stiff as if rheumatic and death follows. To prevent this disease the barges are towed into sea water and moored there for two hours in every fortnight during the summer when temperatures are above 18°C. (Medicines customarily used in Japan are listed in Table 15.)

* The use of *Depterex* was prohibited by law in Japan from 1971, because it is an agricultural medicine.

12. Catching, sorting, transporting and killing eels

ONE MUST ALWAYS CATCH EELS GENTLY BECAUSE DAMAGED SKIN INVITES TROUBLE

Eels have to be caught many times in their lives in order to separate fast and slow growers and move them to larger ponds as they grow, as well as the final harvesting prior to sale. At all stages it is important that the eels should not be damaged. There are three main ways of catching eels in ponds.

1. By draining the pond and catching the eels in a 3 m long net bag tied over the outlet pipe (Fig 72). In case of a bad-drainage pond, the vertical pump in the resting corner (for drainage) is used; the net with bamboo curtain is set in front of the resting corner in order to harvest eels. (Fig 73)
2. By scooping a net around eels gathered at the pond feeding place (Fig 74).
3. By drawing a seine net across the pond (Figs 75, 76, 77).

All these methods give good results and at any given time that which is most appropriate, can be used. To avoid damaging the eels, scoop and drag nets should be hauled very slowly.

The following items must be taken care of:
* Eels are endangered by sudden changes of temperature of 4°C or more, thus if eels are being transferred from one pond to another, the temperature of the receiving pond must be checked in advance and, if necessary, adjusted (by means of changing the water flow) to be similar to that of the donor pond.

* Concentrating eels in a net causes the oxygen content of the water to become much reduced in that area of the pond and the dense mass of eels to gasp for oxygen. To prevent mortality, netting and subsequent size grading of the live eels must be done under conditions of maximum oxygen, thus it is done near the main water inlet in the shallowest part of the pond (in the *resting pool* if there is one), and splasher aerator machines are positioned nearby and run at full speed. If the net is hauled in another part of the pond for some reason, the eels are quickly transferred to a big square keep-net and are moved to the best oxygenated part of the pond as soon as possible. (Figs 78-80, 89)

SORT EELS INTO DIFFERENT SIZES

Captured eels can be sorted into different sizes using mesh containers that allow eels of a certain thickness to escape (this is used chiefly in grading fingerling size, 20-30 cm eels, see Table 16, Fig 88), or by tipping larger eels onto a sloping wooden sorting table whence workmen flick different sized eels into different containers (Figs 71, 81, 82).

STARVE EELS BEFORE SENDING THEM TO MARKET

It takes eels about three days to digest food they have eaten, thus when eels are harvested for marketing they must be kept without food for about three days before they are packed for sending to market. This process is called IKESHIME in Japanese. If this process is not carried out, eels will continue to defecate after packing, their containers will become polluted and the eels will arrive in poor condition, if not dead.

To complete the three-day starvation gut-cleaning period the eels are stored in containers under conditions of plentiful oxygen supply; there are several methods:

* eels are placed in polyethylene perforated tubs under showers in a special shed. The baskets sit in piles on top of each other and the water trickles down from one to the other (Fig 87). A similar method is used in Billingsgate Market, London.
* Eels are placed in polyethylene baskets called *Doman* (Fig 86), about 20 kg eels in each basket, and the baskets are placed in a concrete sluiceway full of fast flowing fresh

68

water (Fig 84), or in the midpond (Fig 85).

After three days the eels will have lost about 5 per cent of their weight due to the complete passing out of their intestinal contents. Starvation beyond this period will cause additional continuing loss of weight, especially if the temperature is warm, which causes the eels to be active (Fig 83). Stored *silver-stage* eels *lose weight much more slowly* than do brown-stage eels and at 10°C can be stored two to three months with little weight loss.

DISPATCHING EELS TO MARKET
Eels are sent to market by one of four methods:
* alive, journeys of more than 8 hours in cool temperatures, in simple boxes packed with a little ice.
* Alive, in double polyethylene bags with oxygen and ice, dispatched by train or truck.
* Alive, in aerated tanker lorries for journeys of up to one week.
* Dead, quick frozen and glazed.

For short journeys in cool climates, such as when sending some eels from Somerset five hours by train to London, eels can simply be put in wooden boxes lined with wet sacking and with a little ice on top. They will arrive in good condition. However, if there is some unexpected delay, they may die.

In Japan, live eels are sent from the culture ponds to Tokyo and Osaka by truck in double polyethylene bags packed with oxygen and ice (Figs 90, 91). Packing procedure is as follows:
1 Take two strong polyethylene bags, put one inside the other.
2 Place 10 kg of live eels in the inner bag.
3 Place some cubes of ice in the bag to lower the metabolic activity of the eels.
4 Inflate the bag with oxygen from a cylinder and tie it.
5 Place bag of eels in a cardboard carton. (Fig 91)

Packed in this way, eels will easily survive for 30 hours and in cool weather or for shorter journeys more eels and less ice can be placed in each bag.

Tanker lorries are used to carry live eels between different parts of Europe. These tanker lorries are huge (Figs 92 to 94). These have been developed chiefly by the eel firm Joh Kuijten of Spaarndam, Netherlands. The most modern lorries weigh 15 tons empty and carry 15 tons of live eels swimming in 15 tons of water. A compressor supplies air bubbles continuously into each tank and the water more or less serves only to keep the eels wet.

Each four days the tanker crew drain the water from the tanks and fill up with new water. In this way they can go for up to 14 days. To bring a load of live eels from Greece to Hamburg is easy. Lough Neagh's eel catch is taken to the continent in 20 hours using such tankers. They even collect eels from the USA: the live eels are loaded into the tanks, the lorry drives to a port, is lifted on the deck of a cargo ship, sits on the deck with compressor running during the voyage, is lifted ashore in Europe and drives to its destination. For a Tanker storage barge see (Fig 93).

Quick-frozen eels are prepared in much the same way as other fish. They are frozen either whole or gutted and cleaned and are quick-frozen either in blocks in a plate-freezer or in an air-blast, and stored at minus 20°C.

Eels have a high fat content and therefore the bodies must be protected from oxidation which will cause rancidity. This is achieved by glazing each eel or block of frozen eels and then wrapping the eel or block in polyethylene and sealing the wrapping. Stored at minus 20°C, such eels will keep in good condition for six months.

KILLING AND CLEANING EELS
There are several ways to kill eels:
1. Put the eels in a deep container, sprinkle on plenty of salt and leave them for two hours. The salt destroys their slime and the eels die of asphyxiation.
2. Put the eels in a tank of freshwater and stun them with an electric shock.
3. Put the eels in trays in a cold store overnight. The eels may not be killed but are slowed down greatly and can easily be handled for gutting etc.

Newly killed eels must be cleaned of slime by washing in cold water and scraping. (A one per cent ammonia solution helps remove slime).

To gut washed eels, the belly is slit open with a knife from the throat to one inch beyond the anus. Using the back of the knife, the guts are scraped out - care is taken to remove the gall bladder without breaking it - and the gutted eel is washed carefully to remove all traces of slime and blood.

Eels can be gutted satisfactorily in quantity by machine. When hand gutting, sawdust or salt sprinkled on the eel enables the worker to get a firm grip on the fish; sometimes a rough cloth is used or the hands are dipped in dry salt for the same purpose. Weight loss during gutting may be 5-10 per cent. Heads are not removed.

Dead eels cannot be deslimed using salt but instead are put in a cold store for a few hours where the formation of ice on the outside of their bodies loosens the slime.

13. Month by month management of ponds in the Hamanako area of Japan

JANUARY

Pond temperatures are low, about 5-8°C, and over-wintering eels are hibernating buried in the mud. Inflow of bore-hole water into ponds is cut off, save for 1-2 hours per day, in order not to disturb the hibernation of the eels by introducing water of a different temperature. No food is given. Blooms of the phytoplankton [Flagellata] may occur during this month. Malachite Green at 0.2 ppm is added to ponds to prevent white fungus disease.

Ponds which have been harvested and elver ponds are dried out at this season and lime is sprinkled on the bottoms to aid the sunshine to disinfect the ponds. The first run of elvers is generally caught in January. They are placed in Nitrofuranaces like *Furanace* (trade name) or *Aivet* (trade name) solution of 1 part per million for one hour to kill any wild bacteria etc. on their bodies.

FEBRUARY

Water temperatures are 3-6°C. This is the coldest month of the year. To prevent marked temperature changes between day and night from disturbing the hibernating eels, water levels in ponds can be raised to make the ponds deeper than usual. Market prices for eels are good at this season. This month being the settlement term of the year, eels are often harvested by draining the pond, but it is dangerous to do so because eels are then subject to injury. The eel's bed should be improved by giving improving chemicals to the bottom soil.

MARCH
Pond temperatures are 10-12°C. The pH value should be kept at 7-8. Blooms of phytoplankton, especially [Scenedesmus], occur as the water warms up for the spring and this causes bad water quality and the pond becomes green. By increasing the rate of water flow through the pond, the transparency of the water should be kept at 15-20 cm. White fungus attacks and other diseases may start this month but as yet there is no effective preventative method. Malachite Green should be sprinkled on the pond, eels closely inspected for disease and any dead eels removed. This is the main month for catching elvers.

APRIL
Pond levels are lowered to speed warming by the sun. Water temperatures are 13-20°C and elvers are fed in their warmed glasshouse tanks. They double their weight in 20 days. In mid-month catch and sort elvers into large and small size groups. Adult eels quit their hibernation and are active and hungry. They become eager to feed when temperatures rise about 18°C but the start of full-scale feeding is usually delayed until this period in order that the ponds shall have attained consistently warm temperatures. Epidemics of white fungus disease often occur this month, simultaneous with the start of feeding. Some antibiotic or sulfa drug added to the feed is believed to reduce the risk of bacteria attacks.

MAY
Water temperature is 20-22°C. Attacks of the white fungus disease die down as the water reaches temperatures in the twenties. Fingerling eels feed vigorously eating artificial feed of about 2.5 per cent, or raw fish of about ten per cent of their body weight per day. Blooms of phytoplankton [Microcystis] bring about good water conditions causing eels to have good appetites. If blooms of water fleas occur the water quality declines and this causes the eels to lose their appetites for the duration of the bloom. *Masoten* should be added to the pond at a concentration of 0.2 parts per million to kill the water fleas and anchor-worm larvae. Salt water is pumped into the pond for 3-4 days to kill anchor-worm young stages. To increase the efficiency of feeding, sort elvers as often as possible (though not necessarily every day) after the first selection.

JUNE

Water temperature reaches 25°C. This is the height of the rainy season in Japan and the weather is hot, humid and much rain falls. In the early mornings, oxygen levels in eel ponds fall dangerously low causing the eels to rise gasping to the surface. The reasons are:

* dense phytoplankton is present and uses up much oxygen at night.
* decomposition of excretion and organic detritus on the pond bottoms use up much oxygen.
* At these high temperatures of 25°C plus, water cannot hold much oxygen in solution.

To provide extra oxygen during the danger hours between 8 pm and 6 am splasher aerator machines in the ponds are switched on and water inflow increased. Daily variations in rainfall and sunshine greatly affect the oxygen level in the ponds and hence the eels' appetites vary from day to day during the hot summer months. Some eels in their second growing season reach marketable 150 gm size this month, and can be harvested with nets.

Swollen intestine disease may occur in this month and can be treated by adding anti-bacterial drugs and vitamin E to the feed. Since oxygen levels are so low at the bottom of ponds, it is vital that no uneaten feed be left in ponds or it will lie on the bottom decomposing anaerobically and further poisoning the water. Scatter lime and ferric oxide (trade name: *Manken*) on ponds to break down organic residues now accumulating on pond bottoms.

Typical daily routine is as follows:

Time	
0600	Collect and analyse water samples from each pond
	Take general look at all ponds
0800 - 1000	Feed adults and elvers
0800 - 1500	Feed elvers (continually)
1000 - 1300	Carry out any regular daily application of chemicals or medicines to ponds
	Remove dead eels
	Declog any filters, outlet sluices

	Do any special tasks of an occasional nature, e.g. catching eels to sell, monthly application of chemicals to ponds
	Maintain equipment and pond
	Sample fish to check growth and health
1500	Skim off green algae that may have grown on surface of covered elver ponds (it can become thick in a few hours sunlight)
1700	Collect and analyse water samples from ponds
1900	Make evening check of all ponds
2000 - 0600	On hot nights check ponds by torchlight and switch on splasher machines if oxygen level runs low and eels are seen gasping at the surface.

JULY

Water temperatures are 27-28°C and the eels feed vigorously. The hot summer is the season when the eels put on most weight. The clarity of the pond water goes through continuous cycles from transparent to turbid, (visibility less than 15 cm) due to successive blooms of phytoplankton. By adjusting the inflow of water, the water quality is maintained in healthy limits. When dense blooms occur, more water is let in and vice versa. Rotifer epidemics may occur. *Masoten* is put in the ponds to kill anchor-worm's larvae.

The Japanese festival called *ushi-no-hi* (day of the ox) occurs this month and everyone likes to eat eel *kabayaki*. Thus many eels 120-150 g are harvested by net to take advantage of the big demand at festival time.

AUGUST

Water temperature is maximum of the year 30-32°C. At this temperature only a little oxygen can dissolve in water and each night the eels have to come to the surface to gasp for breath. To compensate for the lack of oxygen, the splasher-aerator machines are run all day long and maximum volumes of new oxygenated water are admitted to the ponds. Red disease and gill disease occur in the hot temperatures of this month but incidence of swollen intestine disease declines.

SEPTEMBER

Water temperatures are 20-27°C. Heavy feeding of eels

continues. Plankton changes from [Microcystis] to [Oscillatoria] and [Scenedesmus], the water in the pond being green. On some days blooms of [Anabana] occur which greatly reduce the appetites of eels. In ponds containing brackish water, the water turns brown or thick soy sauce colour because of blooms of diatoms, but when the water is settled in colour the eels reveal great appetites again. Splasher aerator machines continue to be run 24 hours a day to keep oxygen levels up. This month is typhoon season in Japan.

OCTOBER
Water temperatures are 18-20°C and the cool of autumn causes the eels' appetites to decrease and vary from day to day. Phytoplankton of a brown colour dominate the ponds. The splasher aerator machines are run each night and sometimes all day too.

NOVEMBER
Water temperatures are 12-15°C. Eels of all sizes gradually lose their appetite. Reduced amounts of feed continue to be given until finally no eels come to the feeding point because they are all in hibernation. Some anti-bacterial drugs and mixed vitamins are mixed with the last week's feed to protect the eels from disease over the winter hibernation. Water depth is increased during winter to afford more protection to the eels. Sprinkle lime and ferric oxide *Manken* on the ponds to make the condition of the bottoms good during the winter.

DECEMBER
Water temperatures are 7-10°C. The eels hibernate except a few may visit shallow edges of the ponds on sunny days. December-March are the main economical months for harvesting eels and this is done by netting or draining ponds. Look back over the year and reflect on your results, then plan for next year. This is important.

14. How to catch elvers

Elvers are the starting point of eel culture. Since eels breed in the sea and since no-one has yet persuaded eels to breed in captivity, eel farmers must obtain stocks of elvers by catching wild elvers that enter river mouths at their accustomed time each year.

The elvers of [A.japonica] and of all the eel species of the temperate zones then enter rivers in huge numbers. Big rivers like the Loire in France receive over something between 50 and 100 million elvers each year.

Elvers do not arrive at all parts of the coast at the same month. For instance in Europe [A.anguilla] elvers enter rivers in Western Spain in December-January, enter the River Severn, Britain in April-May, see Table 17.

There are two reasons for this: first the areas nearest to the spawning ground receive elvers earlier than the remote regions; secondly, elvers are not stimulated to migrate toward river mouths until river temperatures rise above about 8°C. The annual 'wave' of [A.anguilla] elvers arrives at the edge of the European continental shelf in November-December onwards each year, about one year and ten months after being spawned, but the rivers do not become warm enough to attract the elvers until later.

Roughly speaking, elvers of all eel species colonising temperate lands arrive at the coast sometime in the spring. The size of the immigrant elvers varies from species to species. After metamorphising from leptocephalae at sea, elvers do not feed until they enter freshwater, thus they actually lose weight during the period while they wait at sea for rivers to warm up. Elvers entering a river late are thinner

79

and shorter than are the early runs in the same river. To reach Italy, elvers have to travel along the Mediterranean and do not arrive until some months after their brothers have reached Spain. Elvers enter rivers mainly on spring tides, i.e. in a series of waves at two week intervals. They swim only at night and keep right close to the river bank, not in midstream, and enter on each night's rising tide. Usually they swim right at the surface, sometimes in an almost solid procession. Jim Milne, famous elver fishermen of the River Severn, once caught 25 kg of elvers in one scoop of his net. That is about 87,500 individual elvers!! During daylight elvers remain hidden in the mud and under stones. If they come to a weir or waterfall they get around it by wriggling up the damp moss on either side of the falls. To catch elvers there are several methods: see (Figs 95 to 100, 102 & 104)

* Using scoop nets at night at a river bank, to which the elvers are attracted by hanging a bright light.
* Setting a fine-mesh net across the width of a river and so catching all the elvers coming up on a flood tide.
* Constructing a special elver-trap at a weir or waterfall where the elvers upstream migration is obstructed.

A properly designed and built structure can catch almost every single elver that enters a river. On the River Bann in N. Ireland such a system catches an average of some 23,000,000 elvers every year.

Elvers are very delicate. Caught elvers should not be touched by hand but should be placed in a box lined with wet muslin (in the case of small numbers) or hung in mesh cages in the river (large numbers) and then taken to the eel farm within a few hours.

During December-February the coldness of the night air may injure elvers, thus elvers caught at this season should be put in boxes with lids so as not to be exposed to the winter air. By March, temperatures are warmer and elvers are robust and active.

At Epney-on-Severn, England, at St Nazaire on the River Loire and on the River Gironde in France there are special elver catching and storing companies. They buy live elvers from local dip-net fishermen and store the live elvers in tanks of circulating water. Most of the elvers are sent to Germany by aerated tanker trucks (to stock inland waters that receive no natural elver supply) or to Japan by air freight (to stock eel culture ponds). This elver export is big business. The biggest elver exporter and elver expert in Europe is Dr Gerhardt Grunseid of Austria. In the River Bann, Northern Ireland, many elvers are caught but are all put into Lough Neagh.

The annual catch of elvers in these rivers averages:

River Loire	200 tons = 66 million elvers	(many more are not caught)
River Bann	78 tons = 26 million elvers	(virtually all are caught)
River Severn	50 tons = 16 million elvers	(many more are not caught)
River Gironde	50 tons = 16 million elvers	(many more are not caught)

Elvers can be transported by several methods:

* In special aerated tanker lorries (Fig 101). About seventeen tons of brackish water are used to hold one ton of elvers.
* In gauze-bottomed boxes lined with wet muslin or in polyethylene bags with water and oxygen. (Fig 98)
* For air-freight; polystyrene trays (Figs 105-106). The polystyrene is both light and insulates the elvers from sudden temperature changes. Each 0.5 kg tray holds one kg of elvers. The elvers are cooled to 6°C before packing to slow their activity.

The size of the Japanese eel-farm industry and its demand for elvers is now so great that not enough elvers can be caught in Japan itself and agents of Japanese eel co-operatives import many tons of elvers by air-freight from other countries. About 100 tons of elvers and fingerlings are needed each year by Japanese eel farms and they were obtained in 1972 as shown below:

	Elver		Fingerling (below 15cm)	
Japan	10 tons	} [A.japonica]	Taiwan	67 tons
Taiwan[1]	0 tons		Korea	17 tons
Korea[1]	0 tons			
France	64 tons	} [A.anguilla]		84 tons
Italy	2 tons			
England	8 tons			
Total	84 tons			

This requirement for about 100 tons of elvers per year could actually be met by the elver run of a single big river like the

Loire in France. Plenty of elvers arrive from the sea, including river mouths of Japan itself: the problem is how to catch them. The only method to ensure catching *all the elvers* that enter a river is to build a dam right across the river mouth. Usually this is difficult: and expensive involving access for ships, politics, damage to other fisheries etc. But on one big river in N. Ireland it has been done.

A special construction for catching the entire run of elvers is used at the mouth of the River Bann, Coleraine, Northern Ireland. Something like 26,000,000 elvers enter this river every year and every one of them ends up in the two elver traps. At a point about six miles from the sea where the river is about 200 meters wide a dam has been built extending right across the river. The elvers cannot ascend the fast flow of water falling over the central weir and are attracted to two slow-flowing trickles of water coming from the special elver chutes built at either side of the dam. A clever method is here used to trap the elvers. In the special chutes the elvers wriggle up a water-logged 'rope' of straw down which a current of water is trickling. But instead of reaching the upper river, the elvers when they come to the end of the straw drop into a trough. The secret to the working of this device is a slot in the chute which allows some water to fall onto the upper end of the straw rope, thus washing the climbing elvers off the straw into the trough beneath. Elvers always keep to the sides of a river and a loop of the straw rope on the river bed below the trap ensures that all elvers are intercepted.

Table 18 shows the average time needed to grow elvers to market size (200 gms)

[1] Export to Japan of elvers has been prohibited in Taiwan from 1971, and in Korea from 1972.

15. How to cook eels: *Kabayaki*, Smoked and Jellied

KABAYAKI THE JAPANESE STYLE

This is the name of the style in which eels are cooked in Japan, and the food is delicious. *Kabayaki* in Japan, smoked eels in Germany and jellied eels in London are all so tasty that it is impossible to say which is the most delicious.

Kabayaki is prepared by splitting eels, broiling lightly so as to keep the shape while steaming, then steaming the pieces and broiling them over a charcoal fire while periodically dipping them into a tasty sauce called *tare* (Figs 107 to 112). The precise composition of the tare used by each famous chef is secret but the basic composition is a boiled-together mixture of soy sauce: 5 parts, Sweet sake (Mirin): 5 parts and a small amount of sugar.

To prepare kabayaki you take a live eel, pin its head to a wooden board, then slit it up one side of the *back* with one deft cut. Now the body of the eel is opened flat, the guts removed and the backbone is sliced off. The body is cut into pieces about 12cm long and several bamboo skewers or metal needles are threaded through the flesh of each piece. These supports keep the meat flat during the subsequent cooking, preventing it from curling up.

You now lift the first piece on its skewers, and pour boiling water first on the skin side, then on the flesh side. Over steam, the piece is now cooked lightly.

Now you dip the piece in the delicious brown tare sauce and grill it over a charcoal fire. Dip, grill one side, grill other side, dip again etc., about three times, by which time the piece of eel should be an appetising light brown colour (like a

83

kipper). It should also be completely cooked and finally should be giving off mouth-watering smells.

And there you have kabayaki: — eat it in true Japanese style with rice.

Formerly the only places where first grade kabayaki eel was obtainable in Japan were certain famous restaurants specialising in kabayaki. Now, however, kabayaki is also cooked by automatic cookers and can be bought by housewives in supermarkets in vacuum-sealed packs. A number of attractive and savoury recipes have been evolved in the western world for cooking eels and are available in cookery books. Here are given simpler standard methods for general treatment.

SMOKED EELS

Eels are usually hot-smoked after de-sliming and gutting as described in Chapter 12. The smoking method varies considerably from country to country and the method given here is extracted from a helpful pamphlet called *Torry Advisory Note No.37* by J. Horne and K. Birnie, published by Torry Research Station (Ministry of Technology), Torry, Aberdeen, Britain. Kill and clean the eels in the manner described, or use thawed eels from properly frozen stock, and then immerse them in brine for ten minutes; use 270 grm salt to one litre of water.

Thread the brined eels on 0.5 cm diameter rods or spears by pushing the pointed end of the rod through the throat from side to side. The rod may be of stainless steel or wood dowelling. Place small lengths of stick between the belly flaps to keep them apart; this allows smoke to penetrate the belly cavity.

The eels are dried, smoked and cooked during the smoking process. Hang the filled rods in a kiln and smoke the eels for one hour at 35°C (95°F), ½ hour at 50°C (120°F) and finally for one hour at 73°C (170°F). This gradual increase in temperature permits reasonably uniform drying throughout the thickness of the fish; when the temperature is raised too quickly the eels may become case-hardened, particularly when the fat content is low - that is, the skin becomes dry and hard but the flesh remains wet. The eels should lose about 15-20 per cent by weight during the smoking operation.

Remove the eels from the kiln and allow them to cool before packing them; otherwise moulds may form. Brush them lightly with edible oil if necessary and wrap them in transparent plastic film before packing them in boxes.

The finished produce is cooked and ready to eat. The flesh should have a good smoky flavour with only a slight taste of salt; the texture should be firm and buttery, but not too tough. The shelf life of the finished produce is about 3-4 days at chill temperature. 100 lb of fresh, whole eels yields about 60 lb of hot-smoked eels.

SMOKED EELS FOR CANNING
Kill and clean the eels as described earlier and immerse them in 80° brine for 15 minutes. Thread the eels on rods and hang them in the kiln in the same manner as before. Smoke them for one hour at 35°C (95°F) one hour at 50°C (120°F) and finally for one hour at 73°C (170°F). Remove the eels from the kiln, allow them to cool and cut them into pieces the length of the can.

Pack the pieces into cans, fill up with vegetable oil heated to 230°F, and then seal and heat process at 102°C (230°F). A 200 gm (7-oz) oval can takes about one hour. With this method there is very little shrinkage of the meats in the can. For Australian activity see (Fig. 113).

JELLIED EELS
The manufacture of jellied eels (Fig. 114) is a traditional process about which there is little authentic information, and the following hints are given as a guide.

Kill, gut and clean the eels as described earlier, and cut them into pieces about 1½-2 inches long.

The pieces are next cooked by adding them to boiling water, bringing the water back to the boil, adding about 1 lb of salt for every 25 lb of eel, and then simmering until the flesh is soft enough to be pushed off the bone with the fingers. Cooking time will vary depending on the size of the eels, the season of the year and the area of capture. Some experience is required to decide when the eels are properly cooked but about 10 minutes actual boiling is typical for yellow eels. Silver eels have a thicker, tougher skin and require somewhat longer, and they should be overcooked rather than undercooked to make the skin soft enough to eat.

85

Next, cold water is added to bring the oil to the surface. The oil is then skimmed off. The cooked pieces and the hot liquor are next poured into large bowls containing gelatine dissolved in a small amount of hot water, usually about a 10 per cent solution, sometimes with a small amount of added vinegar. The amount of gelatine solution needed depends on the condition of the eels and their natural capacity to gel; again experience is necessary to get the recipe right. Once the mixture has cooled, the pieces in jelly are packed into cartons for fresh consumption; waxed cardboard, plastics and aluminium foil have all been used successfully as containers. Shelf life can be up to two weeks, at chill temperature; waxed carton packs may have a shorter shelf life since there is sometimes some reaction between the gelatine and the wax coating.

Another recipe for jellied eels recommends the use of a salt-vinegar solution and a rather longer cooking time as follows:

Water containing two per cent vinegar and three per cent salt, together with two oz. of spices to the gallon, is brought to the boil and two inch pieces of skinned eel are added. After that addition the mixture is brought back to the boil and then left to simmer for about 45 minutes. The pieces are then put in large bowls to cool and a weak gelatine solution is added if there is insufficient natural jelly.

16. Developments in Japan, Europe, America, New Zealand, Australia and in the tropics

All the main eel culture areas in Japan use still-water culture. The reason is simple. Large volumes of pure water are not available in rivers and, instead, water supplies are obtained from underground bore holes. This bore-hole water cannot be used for running-water ponds: there is not enough of it and it is cold, (only 15-20°C).

In still-water ponds the sun gets the opportunity to heat up the water greatly, even up to 32°C in the height of summer, and thus eels can grow well. By artificially heating the elver ponds starting in March, the number of months per year in which the eels live in water 23°C and above is increased from three to seven months.

The Japanese have tried two further methods of eel culture. Using hot water from the plentiful geothermal hot springs, attempts have been made to warm ponds using this 'free' heat. Unfortunately hot springs do not occur at any of the established eel culture localities and this good idea has not caught on. Culture of eels in sea water has also been tried. The eels grow well, but their taste is reported to be inferior to eels grown in freshwater.

The main new developments occuring in eel culture in Japan are:

* increasing use of "central heating" pipes to warm fingerling ponds during spring time. This increases the length of the annual growing season.
* construction of more and more ponds. Modern ponds are only about 1,000m² in size, much smaller than those built in former years.

* The density of eels in ponds is being continually increased. More and more delicate control of water quality accompanies this practice.

EUROPE Among the countries of Europe many activities concerning eels are in progress. Europe eats 22,000 tons of eels annually and the demand still increases. The main developments are:

* Experimental and commercial attempts to culture eels.
* Research into new techniques of fish culture, especially the use of continuously circulated and purified water in heated insulated tanks.
* Expansion of air-freight export of elvers to Japan. A record quantity of about 75 tons was sent to Japan in 1972 from France, England and Italy.
* Continued research into many aspects of eel biology including a prospective voyage by the Danish research vessel DANA to the Sargasso to attempt to catch the first-ever spawning adult eels and thus complete the research begun by Johannes Schmidt in 1904.

The main problem of eel culture in the eel eating countries of Northern Europe - Germany, Netherlands, Denmark, England - is that they are all too cold. Temperatures of 23°C and over, at which eels grow fast, seldom or never occur (Table 18). In open air ponds eels take four years to reach market size. Japanese eel ponds are in the latitude of Gibraltar and Tunisia.

There are several possible ways to tackle the problem as shown below, and all are being investigated. Which ways will be commercially successful is another question.
1. Use polyethylene multi-span covers to cover still-water ponds, trap solar heat and prevent heat loss due to evaporation and convection (Fig. 115).
2. Build still-water ponds with insulated bottom and sides (panels of styrofoam), covered above with greenhouse sheeting and artificially warm the water.
3. Build running-water ponds, insulated and heated as above, in which the warm water is not discharged but is purified (physically by filters, chemically by resins) and recirculated, thus eliminating heat loss. Such ponds will be small

volume hexagonal or circular structures. Or these ponds could be irrigated by warm water effluent from a power station without recirculation.

4. Quit attempts to culture eels in the cold north and instead develop open-air still-water culture in warmer areas such as around the Mediterranean, perhaps even in the Caribbean.

Work has recently started on the 60 hectare site of the first commercial eel farm in Europe at Sete, south coast of France, financed by a French company - Compagnie des Salins du Midi, known for its production of wine and Mediterranean sea salt. Another French firm in the adjacent Camargue is also reported to be starting a big eel farm.

AMERICA

The USA is the one country in the world where wild eel catches could be greatly increased. Eels occur from Maine to the Gulf of Mexico and are abundant in many areas. Possibly 10,000 tons a year could be harvested but at present only about 600 tons are caught.

Canada has plenty of eels in the St Lawrence which are fully harvested. Due to the cool temperatures, growth rates are slower than in much of Europe.

Eel culture has plenty of academic followers in America but it will probably not develop commercially at present since few Americans like to eat eels.

NEW ZEALAND

A sizeable fishery for wild eels has recently developed in New Zealand and 1,500 tons were caught in 1971. Formerly only Maoris caught and ate the two species of eel in New Zealand and even fought tribal wars over the fishing rights of various areas. Now the eels are caught by many small companies. The eels are nearly all exported frozen to West Germany, Netherlands, Britain and Japan and earn foreign exchange.

Two species of eel occur in New Zealand: the "long-finned eel", [A.dieffenbachi,] which is peculiar to New Zealand, and the "short-finned eel", [A.australis] which also occurs along the eastern coast of Australia. Both species probably spawn near New Caledonia from which area a warm current sweeps down to the Australian coast and round to New Zealand. The elvers reach New Zealand in the spring - September and October - each year.

[A.dieffenbachi] grows to a huge size. Silver females average 6 kg weight and may reach 20 kg. They may stay 30 years in freshwater before migrating to sea; far longer than the European eel (Table 5). The fat content of silver stage specimens is about 25 per cent.

[A.australis] is a modest-sized species that looks very similar to the Japanese and European eels. Female silver eels migrate after about 22 years in freshwater, have a weight of about 600 gm and a fat content of 16 per cent (Table 5).

Near Christchurch on the South Island is one of the phenomenons of the eel world - a brackish lagoon called Lake Ellesmere about 260 km^2 (100 square miles) in size and 2.3 meters deep which is full of eels. An average of 300 tons of eels, containing both species is caught annually.

Lake Ellesmere does not have a permanent connection with the sea but is separated from the sea by a bank of gravel through which water percolates. Elvers can evidently wriggle through the gravel into the lake but silver eels cannot get out. Men catch the silver eels by digging ditches in the gravel from the lake towards the sea during the silver migration season. The silver eels detect the increased flow of water out of the lake and at night swim in huge numbers into the ditches in an attempt to get to the sea; but instead they are caught. (Similar lagoons exist in Newfoundland, Canada).

To encourage the eel industry, efforts are being made by the Fishing Industry Board of New Zealand (general manager Mr. J. S. Campbell) and Victoria University, Wellington Dr. P. H. J. Castle, Zoology Department).
The main objectives are:

* To expand the catch of wild eels to the maximum sustainable yield, but not beyond.
* To encourage development of eel culture.
* To do research to discover more about the biology of New Zealand eels.

Eel culture will probably get best results using the still-water method and will be located in the northern parts of North Island around Hamilton, Auckland and the Northland where water temperatures are warmest: temperatures of 23°C and above occur during two months each year (Table 18).

Gratitude is due to Mr. J. S. Campbell and Mr. P. Chapman of

90

New Zealand Fishing Industry Board and Dr. P. H. J. Castle, Mr. P. R. Todd and Mr. D. J. Jellyman of Victoria University Zoology Department for information and data supplied.

AUSTRALIA Two species of eel occur along the coasts of Eastern Australia and commercial quantities are found in Victoria, Tasmania and New South Wales. About 150 tons per year are caught, mainly by fyke nets and there is interest in the possibility of starting eel culture. A State licence to start an eel farm in Tasmania was granted early in 1973.

[A. australis] is the most important species and occurs from western Victoria north up to Brisbane. It is the same species which occurs in New Zealand. It is important because it is the most abundant species (about 200 tons caught/year) and also because in size and plain colour it closely resembles the European and Japanese eel. The greatest biomass live in Victoria and nearly all Australia's catch is taken here at present. The eels are exported frozen to Europe.

[A. reinhardti] is a big mottled species and occurs from Tasmania right up to northern Cape York and also on New Caledonia. It grows to a large size, is presently used only for shark bait and only about ten tons are caught annually.

Both species probably spawn to the east of New Caledonia. Elvers enter rivers in Queensland in winter and in New South Wales and Victoria in spring as listed below:

Queensland:	elvers enter rivers June, *July, August,* September
New South Wales:	elvers enter rivers July, *August, September,* October
Victoria:	elvers enter rivers August, *September, October,* November

These records refer to unidentified elvers, since as yet no research biologist has carried out the painstaking task of separating in each sample the elvers of the two different species.

[A. australis] reach silver stage at about 600 gm, the same as in New Zealand and [A. reinhardti] silver eels are very large, about 5,000 gms (silver-stage sizes are given in Table 5, maximum sizes are given in Table 4).

It seems probable that the wild population of eels in Australia is not high and that wild eel harvests cannot be increased dramatically. The climate of Queensland is fully as warm as Japan and would be suitable for eel culture (Table 18).

91

Thanks are due to Mr. P. C. Pownall, editor of *Australian Fisheries* and Dr. D. Buckminster of Victoria Freshwater Fisheries Department, for data supplied.

TROPICAL COUNTRIES

It is in tropical countries that the majority of [Anguilla] species live (Fig. 6). Indonesia is the ancestral home of the genus and no fewer than nine species live there to this day. Other species live in East Africa, Madagascar, India, Burma, New Guinea and the Pacific Islands.

It is not believed that large catches of wild eels can be taken in these tropical areas because although many individual eels occur, some very large, the total biomass is small.

Culture of eels is also unlikely to develop for the simple reason that the people in tropical countries need all the protein they can get. They cannot afford the luxury of raising carnivorous animals which destroy protein. Producing one kg of eel flesh requires that the eels eat about 7 kg of the flesh of some cheaper fish.

What is wide open to be tackled in tropical countries is biological research into the lives of the tropical eel species. The biologist who does this has all the discoveries listed in Chapter 3 open to him and will become a famous man. As yet hardly anything is known about the lives of the tropical eel species.

17. Check list of requirements for starting an eel farm

1. CHARACTER OF MAN IN CHARGE
a) Do you have a sound business sense and knowledge of financial procedures? Do you understand the scientific details of eel biology?

b) Are you of a quiet temperament and manually skilful at practical engineering such as making cement structures? If you are a high-powered tycoon-type you will quickly get bored with eel farming and quit.

2. CAPITAL FINANCING
a) Have you sufficient capital to cover the capital costs of starting a farm?

b) Have you in addition sufficient capital to pay the running costs of the farm?

3. PRELIMINARY ENQUIRIES
a) Have you read all the available literature on eel culture? You should make visits to commercial fish farms and government culture research stations to see what they look like and meet people who can advise you. Can you afford a two-week trip to Japan itself to see the real eel culture industry?

b) Have you sought the advice of experts in the following fields: running a business, banking, accountancy, real estate, engineering, biology?

c) To set up an eel farm you will need permission from your local river board and county planning authority. These people have much helpful advice to give: the earlier you contact them the more they can help you.

d) Make out a detailed costing of your operation. Estimate costs high and production low.

4. PHYSICAL FEATURES OF SITE

a) Water supply volume. Is at least 40 million gallons (18,000m^3) available in the driest months of the year? It can come from a river, lake, as well as borehole.

b) Water supply quality.

1) Is the water supply free from possibility of dangerous pollution? Check the whole upstream part of any river you propose to use to see if there are any factories and sewage outlets which may cause trouble. Ask farmers what insecticides they use and when. Water need not be crystal clear - murky water is fine. If wild eels occur in the water, that is a proof it is of good quality.

2) How can you detect pollutants entering your farm?

3) How can you prevent pollutants entering your ponds?

4) What recourse have you in case of loss due to pollution?

5) Can technical assistance be obtained for detection of pollution and at what cost?

c) Flooding: Are you sure that the site can never be flooded?

d) Area: You need about 4 acres (1½ hectares) to build a 40 ton/year production farm and 10 acres (4 hectares) more for future expansion. (See Table 9) Can you obtain this land by purchase or by lease?

e) Services: Road access and electricity are vital. A telephone is handy.

5. BIOLOGICAL FEATURES

a) Elvers. From where will you get your elvers? At what cost In which months are they available?

b) Diseases. If eels become diseased, what facilities have you for diagnosing which disease is responsible? Where is the nearest parasitologist/fish disease expert who can help you? What costs may be involved?

c) Feeding: What is the present price per ton of artificial feed? What is the price of raw fish and how much per ton does it cost to (a) freeze and (b) store raw fish?

6. MARKETING

a) Your profit largely depends on the price at which you sell your eels. You need marketing know-how and skill. Will you sell your eels in bulk at a low price to a firm that will collect them? Or will you supply smaller amounts direct to retailers at a higher price? Which firms have you contacted?

b) Do you want to smoke your eels yourself and pack them?

These and many other questions must be considered before undertaking the venture but, if properly based, there is ample evidence to indicate that the steadily rising price of eels because of ever expanding demand offers a growing attraction for profitable investment and enterprise.

18. Useful books and magazines

A peculiar situation exists regarding literature on fish culture due to the fact that most persons actively operating farms are in the business for profit and do not wish to reveal their trade secrets. Therefore books tend to be written by biologists and academic people and tend to be either vague, academic or out-of-date. Useful scientific and commercial information is none the less to be found in print, and more will be published as the years go by.

A huge number of articles on aspects of aquaculture have been published in Japanese and dozens more are published each year. They are published in Japanese because they are intended for local use and are mostly produced by the prefectural experimental stations. Several good textbooks exist, and recently several practical laymans guides to commercial culture of certain species have been published; books on Yellowtail and Eel culture are available so far. In the following list, the English translation of each title is given first.

This * marks publications of particular practical use to eel farmers.

Books and articles
* *Farming the Edge of the Sea* (survey of coastal culture) by E. S. Iversen, 1968. Fishing News (Books) Ltd., 110 Fleet Street, London E.C.4 and 23 Rosemount Avenue, West Byfleet, Surrey, England.
Textbook of Fish Culture (comprehensive study) by Marcie Huet, 1973. Fishing News (Books) Ltd., 23 Rosemount

Avenue, West Byfleet, Surrey, England.

Danish Eel Investigations during 25 years, 1905-1930 by J. Schmidt, 1932. Carlsberg Foundation, Copenhagen, Denmark.

A revision of the genus Anguilla by V. Ege, 1939. Carlsberg Foundation, Copenhagen, Denmark.

* *Eels - a biological study* by L. Bertin, 1956. London, Cleaver-Hume Press.

* *Eels: How to catch them* by Raymond Perrett, London, Barry and Jenkins.

* Fish and Shell Fish Farming in Coastal Waters, 1972. P. H. Milne, Fishing News (Books) Ltd., 23 Rosemount Avenue, West Byfleet, Surrey, England.

* *Lessons in fish farming from Japan* by G. R. Williamson, 1971, in Fishing News International, published by Arthur J. Heighway, Publications Ltd., 110 Fleet Street, London E.C.4. May and June 1971 issues.

* *Aquariums* by Anthony Evans, London; Foyles.

Elements of Aquaculture methods Yogyo gaku so ron (in Japanese) edited by N. Kawamoto, 1965. Pub. by Kosei-sha Kosei-kaku, Sanei-cho 8, Shinjuku-ku Tokyo.

Aquaculture methods for all species. Yogyo gaku kaku ron (Japanese) ed. N. Kawamoto, 1967. Pub. as above.

* *Eel.* Unagi (in Japanese) - a practical guide to eel culture by Sanya Iizuka 1971. Pub. by Nosan Gyoson Bunka Kyokai, 7-6-1 Akasaka, Minato-ku, Tokyo.

* *Eel Study.* Man Gaku (in Japanese). Vol. 1: biological study, Vol. 2: technical study on culture. By Isao Matsui 1972, Published by Kosei-sha Kosei-Kaku, Sanei-cho 8, Shinjuku-ku, Tokyo.

* *Eel.* Unagi (in Japanese); a review of eel culture by Shun Inaba, Isao Matsui, Hideaki Tsunogai, Hiroshi Aoe and Hirohisa Ogami, 1971. Pub. by Midori Shobo, Fuji Bldg., 4-6-5 Iidabashi, Chiyoda-ku, Tokyo.

* *A Book about Eel,* Unagi no hon (in Japanese) containing essays, legends and miscellaneous writings about eel by Isao Matsui, 1971. Pub. by Marunouchi Shuppan, 5 Maru Bldg., Chiyoda-ku, Tokyo.

* *Theory and Practice of Eel Culture,* Yoman no riron to jitssai (in Japanese) by Isao Matsui 1970. Revised edition. Pub. by Nippon Suisan Shigen Hogo Kyokai, 1-11-35 Nagata-cho, Chiyoda-ku, Tokyo.

Study of Freshwater Propagation, Tansui zoshoku gaku (in Japanese) by Densabro Inaba 1966, Pub. by Kosei-sha Kosei-kaku.

Magazines and Journals

* *Fish Farming International,* 1. (For current developments in fish culture the world over. To be followed by successive volumes as needed.) Fishing News (Books) Ltd., 23 Rosemount Avenue, West Byfleet, Surrey, England.

  FAO Fish Culture Bulletin, pub. by Fisheries Division, Biology Branch, FAO, Via delle Terme di Caracalla, Rome.

  The Canadian Fish Culturist, pub. by Dept. of Fisheries of Canada, Ottawa, Canada.

* *The Progressive Fish Culturist,* pub. by Superintendent of Documents, U.S. Government, Printing Office, Washington D.C. 20402, USA.

  The Fish Farmer, pub. by 1378 Livermore Avenue, Livermore, California 94550, USA.

  American Fish Farmer, published by P.O. Box 1900, Little Rock, Arkansas 72203, USA.

* *Farmers Weekly* (for contacting landowners and for advertisements of construction products etc.) pub. by Farmers Weekly, 161 Fleet Street, London E.C.4.

* *Grower* (for ideas from horticulture industry and for advertisements) pub. by Grower Publications Ltd., 49 Doughty Street, London W.C.1.

* *Electrical Surplus* (for advertisements of second-hand pumps etc.) pub. by R. F. Winder Ltd., Belgrave Electrical Works, Stanningley, Leeds. Tel. Pudsey 77621.

* *Fish Culture* Yoshoku (in Japanese, monthly trade magazine) Pub. by Midori Shobo, Tokyo.

19. Meanings of some Japanese words

unagi	eel
shirasu unagi	elver
byoki	disease
doman	round container for holding eels
kenkyusho	research laboratory
kumiai	co-operative society
ikeshime	method of starving eels to clean their guts
sakana	fish
shikenjo	experimental station
suisan	fishing industry
yoshoku	fish culture
yoman	eel culture
yoshokujo	fish farm
sanso	oxygen
jinko jiryo	artificial feed
- cho	sub-section of a town
- ken	prefecture
- ku	area of a city (ward)
Suisancho	Fisheries Agency, Ministry of Agriculture and Forestry, Japanese Government

20. Status and organization of aquaculture in Japan

Aquaculture is an important and growing industry in Japan. In 1970, the annual total production of fisheries was 9,315,000 tons in volume comprising 549,000 tons of shallow sea culture, and 48,000 tons of inland water culture, nine hundred million yen in value, indicated according to species of fish. Of this eel culture amounted to 19,456 tons, valued at 220.8 hundred million yen.

Fry supply, research, legislation and subsidies all contribute to the aquaculture industry and many firms make equipment specially designed for aquaculture.

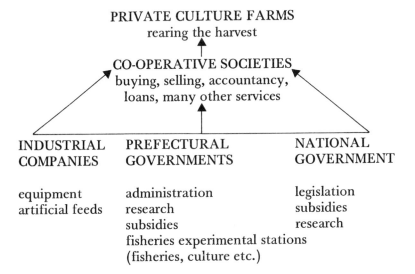

PRIVATE CULTURE FARMS
rearing the harvest

CO-OPERATIVE SOCIETIES
buying, selling, accountancy, loans, many other services

INDUSTRIAL COMPANIES	PREFECTURAL GOVERNMENTS	NATIONAL GOVERNMENT
equipment artificial feeds	administration research subsidies fisheries experimental stations (fisheries, culture etc.)	legislation subsidies research

At the present time, and for several decades past, a feature of

Japan has been the very heavy expenditure of public funds on fisheries. Part of this money is used to support aquaculture, although the majority is devoted to normal fisheries. There are at the present time in Japan:

> 79 main and 76 branch fisheries experimental stations (including breeding centres and culture experimental stations).
> 17 fisheries universities and colleges.
> 1,500 staff members (including biologists, research chemists and technicians) and 457 advisors engaged in guidance to fishermen and fish farmers in Prefectural Government employ.

Undoubtedly this massive Government backing has helped the industry greatly in the past. Now, however, the industry seems to make progress chiefly by its own efforts. The best aquaculture men are (1) farmers themselves, (2) the staffs of breeding centres and culture experimental stations and (3) excellent biologists at experimental stations and university fisheries faculties.

PRODUCTION
The marketed harvest is mostly produced by many small companies or individual farmers. These are usually members of co-operative societies. The co-operative societies are important and serve many functions; to market the produce of members, to purchase fry in bulk for members, to provide accountancy services, to act as the farmers mouthpiece when dealing with government etc. Only a few large companies engage in fish farming.

FRY SUPPLY
Fry of rainbow trout, carp etc., seed of abalone, scallop etc. and young prawns are produced artificially at 37 special breeding centres, since they cannot be caught wild in adequate numbers. To operate a breeding centre is a sophisticated and expensive business, therefore these centres are operated by the Prefectural Governments and have scientific staffs. From the centres the fry are sold at subsidised prices to co-operatives or individual farmers. Young invertebrates and fry such as prawns, blue crabs and rock fish etc. are

raised in National Government Farming Centres and are subsequently released directly into the Inland Sea to increase wild stocks of these species.

The breeding centres are of key importance and as the list of cultured species expands, their work will increase. They also serve as demonstration farms where private fish farmers and others can see the latest techniques and equipment in use.

RESEARCH

About 155 prefectural experimental stations exist scattered all over Japan, sometimes more than three in a prefecture. At 137 of these, aquaculture is among the subjects studied. The aim of these stations is practical and local, and many are doing almost identical work. These stations are in contact with the co-operative societies of fish farmers in their areas and can arrange visits to individual farms. Culture research of a more academic and usually less practical nature is carried out at the two fisheries universities and 15 university fisheries faculties. The *Suisancho* (Fisheries Agency) operates eight fisheries research laboratories, six of them doing culture research.

Free advice for farmers is available at any of the 137 prefectural fisheries experimental stations and six fisheries research laboratories of *Suisancho* and 17 fisheries universities and this is an important aid. Research reports in Japanese on aquaculture are produced regularly by most of these stations, laboratories and universities.

A huge amount of information on aquaculture has been published - but largely in Japanese.

ADMINISTRATION

The *Suisancho* (Fisheries Agency, Ministry of Agriculture and Forestry) in Tokyo is the body which administers aquaculture at a national level, and consists of Fishery Administration Department, Oceanic Fishery Department, Fishing Port Department and Research and Development Department to provide administrative guidance, promotion of subsidies and research study. This Research and Development Department has a staff of 403 researchers, including 199 engaged in culture research.

105

21. Suppliers of Feeds and Equipment in Japan

To service the needs of fish and eel farmers, specialities are provided by a number of reliable firms, as follows:

ARTIFICIAL FEEDS FOR MANY FISH SPECIES
Nippon Formula Feed Mg. Co., Ltd., 2-9, Nishi-shinbashi 1-chome, Minato-ku, Tokyo.
Chubu Shiryo Co., Ltd., 2-96, Tsutsumi-cho, Minami-ku, Nagoya-city, Aichi prefecture.
Nisshin Flour Milling Co., 1-2-4, Koami-cho, Nihonbashi, Chuo-ku, Tokyo.
Nihon Nosan Kogyo K.K., 2-2, Shinurashima-cho, Kanagawa-ku, Yokohama-city, Kanagawa prefecture.
Yeaster Co., Ltd., 726, Fukuda, Honda-cho, Tatsuno-city, Hyogo prefecture.
Oriental Yeast Co., Ltd., 6 Azusawa 3-chome, Itabashi-ku, Tokyo.
Feed Department, Taiyo Gyogyo K.K., 1-4, Marunouchi, Chiyoda-ku, Tokyo.

FEED OILS TO ADD TO ARTIFICIAL FEEDS
Riken Vitamin Oil Co., 3-8-10, Nishi-kand, Chiyoda-ku, Tokyo.
Nippon Chemical Feed Co., Ltd., 3-6, Asano-cho, Hakodate-city, Hokkaido.
Nippon Vitamin Oil Industrial Co., Ltd., 155, Nakamaruko, Nakahara-ku, Kawasaki-city, Kanagawa pref.

NETS, FISHING GEAR, FISHERIES EQUIPMENT etc.
Nichimo Co., Ltd., 6-2, Otemachi, Chuo-ku, Tokyo.

Taito Seiko Co., Ltd., 1-2, Higashi-shinbashi 1-chome, Minato-ku, Tokyo.

FLEXIBLE TARPAULIN TANKS AS CULTURE PONDS
Toray Industrial Inc., 2-2, Muromachi, Nihonbashi, Chuo-ku, Tokyo.
Asahi Chemical Industrial Co., Ltd., 1-12, Yurakucho, Chiyoda-ku, Tokyo.
Unitika Ltd., 7-20, Yaesu 1-chome, Chuo-ku, Tokyo.

STC POLYCARBONATE TANKS FOR REARING LARVAE AND ROTIFERS
Teijin Chemicals Ltd., 6-21, Nishi-shinbashi 1-chome, Minato-ku, Tokyo.

STYROFOAM BARREL-SIZE FLOATS FOR NET CAGES
Sekisui Plastics Co., Ltd., 2, Kinugasacho, Kita-ku, Osaka.
Tokyo Cork Co., Ltd., 1-18, 5-chome, Osu, Hiroshima-city, Hiroshima prefecture.

VERTICAL PUMPS AND PADDLE SPLASHERS FOR POND-CURRENT/AERATION
Yasuda Denki Kogyosho K.K., 4, Nishikicho, Toyohashi-city, Aichi prefecture.
Hamamatsu Denko Sha Co., Ltd., 1958, Tennomachi, Hamamatsu-city, Shizuoka prefecture.

COMPLETE KITS FOR MAKING GREENHOUSES OF POLYETHYLENE SHEETING STRETCHED OVER FRAMES OF STEEL TUBES
Mitsui & Co., Ltd., 2-9, Nishi-shinbashi 1-chome, Minato-ku, Tokyo.
Tsukiboshi Kogyo Co., Ltd., 11-5, 4-chome Hatchobori, Chuo-ku, Tokyo.

TABLE 1: World production of eels per year

Average of 1968-71 data from FAO and other sources. World annual production is about 25,000 tons of wild eels and 26,000 tons of cultured eels. Consumption rating: *** high, ** medium, * low. The USA is the only area of the world where wild eel catches could still be significantly increased.

Area/Country	Species	Wt. caught/year metric tons	Consumption rating	Net importer or exporter
Europe and N. Africa	[A.anguilla]			Imported
Denmark	[A.anguilla]	3,400	***	imported
Italy	[A.anguilla]	3,400	*	exported
Netherlands	[A.anguilla]	2,800	***	imported
France	[A.anguilla]	2,500		exported
Sweden	[A.anguilla]	1,900	**	balanced
West Germany	[A.anguilla]	1,500	***	imported
Spain	[A.anguilla]	1,400		exporter
East Germany	[A.anguilla]	1,300	**	balanced
Poland	[A.anguilla]	1,000	*	balanced
N. Ireland and Eire	[A.anguilla]	900		exported
USSR	[A.anguilla]	600	*	balanced
Norway	[A.anguilla]	600		exported
Morocco	[A.anguilla]	300		exported
England, Wales, Scotland	[A.anguilla]	100	*	imported
Belgium	[A.anguilla]	100	*	imported
Tunisia	[A.anguilla]	100		exported
Greece	[A.anguilla]	100		exported

Table 1 continued

Area/Country	Species	Wt. caught/year metric tons	Consumption rating	Net importer or exporter
North America	[A.rostrata]			Exported
Canada		1,000		exporter
USA				balanced
Asia	[A.japonica] (mainly)			
Japan	wild	3,000	***	balanced
	cultured	24,000		
Taiwan	cultured	2,000	*	balanced
S. Korea	cultured	200	*	balanced
China	wild	6,000	**	balanced
Australia	[A.australis]	200		exported
New Zealand	[A.australis] [A.dieffenbachi]	1,000		exported

TABLE 2: Waters yielding heaviest sustained catches of wild eels

Body of water	eel catch/yr average Metric tons	area km^2	sq. miles
Ijssel Meer, Holland	2,300	1,800	700
Lough Neagh, N. Ireland	800	400	153
Commachio area, Italy	680	325	125
Lake Ellesmere, New Zealand, S. Island	300	260	100

TABLE 3: Main eel eating countries and their preferences

Ranking	Country	Preferred size & type of eel		style eaten	Approx. Wholesale price
		Wt. range gm	silver or brown		£/kg 1971
1	Japan	150-200	cultured and wild	*Kabayaki*	1.75 (cf. Y 1,508/kg) 1 L = Y 860 1971
2	W. Germany	over 260	silver	smoked	1.10
3	China	120-180	brown or silver		
4	Netherlands	50-260	brown	smoked	0.65
5	Denmark	300-400	brown or silver	smoked	1.10
6	Sweden	over 450	silver	smoked (sliced)	1.30
7	E. Germany	over 260	silver	smoked	0.90
—	Britain	110-350	silver or brown	cold jellied or hot stewed	

cf.) Wholesale prices of eels in Japan in 1971 were as follows:

Jan.	Y 1,368/kg	Jul.	Y 1,603/kg
Feb.	Y 1,611/kg	Aug.	Y 1,524/kg
Mar.	Y 1,642/kg	Sep.	Y 1,347/kg
Apr.	Y 1,757/kg	Oct.	Y 1,247/kg
May	Y 1,520/kg	Nov.	Y 1,318/kg
Jun.	Y 1,468/kg	Dec.	Y 1,354/kg

The prices of eels in Japan are not so low in Summer as in Europe because demand is increased in that season.

TABLE 4: Details of world's species of *Anguilla* eels
Based on Ege (1939) Size data from various reports and field workers.

Colour of Body	Name	Number of vertebrae (counting hypural as being the last vertebra) *Average*	Distribution	Approx. max. size of female eels kg	cm	Notes
mottled	[A.ancestralis]	103	N. Sulawisi			
mottled	[A.celebesensis]	103	Indonesia, Philippines			
mottled	[A.interioris]	105	New Guinea			
mottled	[A.megastoma]	112	Pacific islands from Solomons east to Pitcairn	22	190	
mottled	[A.nebulosa]	110	East Africa and India	10	150	
mottled	[A.marmorata]	106	South Africa, Madagascar, Indonesia, China, Japan, Pacific Islands	27	200	[A.marmorata] is the most widely distributed species
mottled	[A.reinhardtii]	108	Eastern Australia, New Caledonia	18	170	
plain	[A.borneensis]	106	Borneo, Celebes	2	90	
plain	[A.japonica]	116	Japan, China	6	125	[A.japonica], [A.rostrata]
plain	[A.rostrata]	107	E. coasts of USA, Canada, Greenland	6	125	and [A.anguilla] are
plain	[A.anguilla]	115	W. coast of Europe, N. Africa, Iceland	6	125	closely related
plain	[A.dieffenbachii]	113	New Zealand	20	150	
plain	[A.mossambica]	103	South & East Africa, Madagascar	5	125	
plain	[A.bicolor]	108	E. Africa, Madagascar, India, Indonesia, N. W. Australia	3	110	
plain	[A.obscura]	104	New Guinea, Pacific islands from Solomons east to Tahiti			
plain	[A.australis]	112	Eastern Australia & New Zealand	2.5	95	

TABLE 5: Size, age and period of migration of silver-stage eels of various species

The figures given are the average values. Note that male silver eels return to the sea at much smaller sizes and about a month earlier than female silvers. Each species of eel returns to the sea when it reaches a certain size: the number of years taken to reach this size varies according to the growth rate in each locality which in turn is determined chiefly by the water temperature and amount of food available.

Species	Locality	Female silver eels				Male silver eels			
		wt	length	age	season of migration	wt	length	age	season of migration
[A.anguilla]	N. Ireland	260 gm	50 cm	12 yrs	Sept-Oct	120 gm	40 cm	9 yrs	August
[A.rostrata]	Canada, St Lawrence river,	1,600 gm	91 cm		Sept-Oct				
[A.rostrata]	Newfoundland, Topsail Pond	600 gm	70 cm	12 yrs	Sept				July-August
[A.japonica]	Japan, Southern Honshu	200-250 gm	58 cm	7 yrs	Sept-Oct	100-150 gm	35 cm	5 yrs	August
[A.japonica]	China, West River	270 gm	55 cm	3 yrs	Oct-Nov	140 gm	46 cm	3 yrs	Sept-Oct
[A.marmorata]	China, West River	12,000 gm	150 cm	12 yrs	Oct-Jan	1,500 gm	80 cm	8 yrs	Sept-Nov
[A.dieffenbachi]	New Zealand, Lake Ellesmere	6,000 gm	120 cm	30 yrs	May	600 gm	64 cm	20 yrs	April
[A.australis]	New Zealand, Lake Ellesmere	600 gm	68 cm	22 yrs	April	200 gm	46 cm	14 yrs	March
[A.australis]	Australia, Victoria, Lake Yambuk	600 gm	70 cm		Feb-March	200 gm	45 cm		Jan-Feb.
[A.reinhardti]	Australia, Victoria	5,000 gm	110 cm		March-April				Feb-March

TABLE 6: Details of ponds used to culture eels in Japan

Size of eels grown in pond	Approx. area of pond	Depth	Bottom	Covered or open	Temperature	Type of water supply	Notes	Approx. period in use
New elvers	20m²	0.6m	concrete	under greenhouse	heated to above 25°C	running-water	aerated by bubblers	March
Elvers 8-12 cm	30-100m²	1m	concrete	under greenhouse	heated to above 25°C	running-water	aerated by bubblers	April-May
Fingerlings 12-20 cm	200-300m²	1m	mud	open	natural	still-water	overhanging lip to pond. Green water	June-July
Adults 20-70 cm	500-1,000m²	1m	mud	open	natural	still-water	green water	August to harvest in the autumn of following year

TABLE 7: Weight of eels per m^2 of bottom that can live in still-water mud-bottom ponds

These values apply to still-water mud-bottom ponds in which the eels burrow in winter to hibernate. Future running-water concrete-bottom artificially-heated ponds in which eels grow continuously may be stockable at higher densities. In this case, maximum oxygen must be supplied by using paddle splasher machine etc.

Pond size m^2	kg eels/m^2 sustainable
10,000	1.0
500	2.5
200	4.0*

TABLE 8: Monthly weight increases of cultured eels during their first growing season in Japan

Date	Per cent of year's weight increase	Average temperature of water $^{\circ}C$
End of March (new caught elvers)	0	11
End of April	1	16
End of May	3	21
End of June	7	25
End of July	14	28
End of August	24	30
End of September	24	23
End of October	22	19
End of November	4	14
End of December	0	9
TOTAL	100	

TABLE 9: Pond space required to yield 20 tons market size eels per year. 50 kg of elvers are needed to produce 20 tons market size eels.
A water supply of about 450 m³ (100,000 gallon) per day is needed.

size of eel stocked	weight stocked		weight & length of average eels		stocking density		Pond area needed m²	Number and size of ponds
	start	finish	start	finish	start	finish		
elvers	50 kg	150 kg	0.16 gm 6 cm	0.5 gm 8 cm	0.4 kg/m²	1.2 kg/m²	125m²	6 ponds 20m² (5m diameter) size, running water
older elvers	150 kg	400 kg	0.5 gm 8 cm	1.3 gm 12 cm	0.5 kg/m²	1.6 kg/m²	300m²	8 ponds 40m² (8 x 5m) size running water
fingerlings	0.4 tons	2 tons	1.3 gm 12 cm	6.5 gm 20 cm	0.4 kg/m²	2.0 kg/m²	800m²	10 ponds 80m² (10 x 8m) size still water
adult eels	2 tons	20 tons	6.5 gm 20 cm	190 gm 70 cm	0.4 kg/m²	4.0 kg/m²	5000m²	25 ponds 200m² (14 x 14m) size still water

TABLE 10: Cost of constructing eel ponds in Japan 1967
Prices in Yen, 850 Yen = £1 (present 1973 rate: Y 660 = £1)

	5,000m^2 pond lined with stones and cement	12,000 m^2 pond lined with concrete slats
Work of bulldozer to excavate pond	1,140,000	1,600,000
Stones and cement	400,000	
Concrete slats and posts		600,000
Electrical wiring	250,000	300,000
Digging well and laying water distribution pipes	500,000	450,000
Vertical pumps	60,000	180,000
	(1 pump)	(3 pumps)
Paddle splasher machines	60,000	240,000
	(1 splasher)	(4 splashers)
Feed Store	400,000	450,000
	Y 2,810,000	Y 3,820,000
	£ 3,300	£ 4,500

TABLE 11: Conversion ratio of artificial feed and raw fish

Type of Feed	Conversion ratio	Meaning
Artificial feed	1.4:1	1.4 kg of feed (dry weight) must be eaten by the eels to bring about a weight increase of one kg.
Raw fish	7:1	7 kg of feed must be eaten by the eels to bring about a weight increase of one kg.

TABLE 12: Composition of raw fish

Species	per cent composition			
	Water	Protein	Fat	Carbohydrate
Mackerel [Scomber]	76	18	4	0.7
Atka mackerel [Pleurogrammus]	77	17	5	0.2
Pacific Saury [Cololabis]	70	20	8.4	0.2
Sardine [Sardinops]	75	17	6	0.8
Horse mackerel [Trachurus]	75	20	3	0.7
Eel [Anguilla]	61	20	18	0.3

TABLE 13: Monthly consumption of food on a typical eel farm

	Per cent of weight of food eaten annually	Average temperature of water °C
January	0	7
February	0	5
March	1	11
April	2	16
May	3	21
June	7	25
July	15	28
August	24	30
September	28	23
October	18	19
November	2	14
December	0	9
TOTAL	100	

TABLE 14: Diseases of Eels

Name	Japanese name	Illustration	Symptoms	Cause	Cure
Fungus disease (Saprolegniasis)	watakamuri-byo	Fig.62	white fungal growths on outside of body	fungus [Saprolegnia parasitica]	add anti-bacterial drugs to feed put Malachite green or Methylene blue in pond water
Red disease	hireaka-byo	Fig.68	red colour on fins and skin	[Aeromonas liquefacines]	add anti-bacterial drugs to feed
Swollen intestine disease	choman-byo		intestines swollen	bacterial infection	add anti-bacterial drugs to feed
Gill disease	era-byo	Fig.69	gills raw and rotten	bacterium [Chondrococcus columnaris]	add anti-bacterial drugs to feed
White spot disease (Ichthyophthiriasis)	hakuten-byo	Fig.67	tiny white spots on outside of body	[Ichthyophthirius multifiliis]	pump salt water into pond
Myxidium disease	myxidium-byo		round white expanse of cysts with spores on skin	protozoan [Myxidium]	unknown
Cripple body disease (Plistophorasis)	beko-byo	Fig.65	body kinked or misshapen	protozoan [Plistophora anguillarum]	unknown (remove dead eels)
Gas disease	kiho-byo	Fig.66	bubble on head (elver) or fin (adult)	too much O_2 or N_2	add new water to pond or remove eels to better condition pond
Anchor worm disease	ikarimushi-byo	Fig.70	parasitic crustacea attached inside mouth of eel	parasitic copepod [Lernaea cyprinacea]	[Masoten] and bleaching powder kill larvae by sprinkling on pond: kill eggs and larvae by pumping salt water into pond
Branchionephritis	erajin-en		gills swollen, raw, rotten etc. and kidney inflammation	unknown	add salt to pond (0.5-0.7% in salinity)

TABLE 16: Mesh size of wire mesh containers for grading fingerling eels, and the size of eels which they permit to escape

Mesh size	Size of eel that can escape the mesh
7 mm	7 gm
9 mm	13 gm
11 mm	19 gm
13 mm	37 gm

TABLE 17: season of elvers run in parts of Europe

Place	age of elvers entering rivers (Spawning is in February)		Month of main elver run	Size of elvers (no. per one Kilo)
	Years	Months		
Spain and Portugal, atlantic coasts	1	10	Dec-Jan	2700
France, Biscay and Brittany coasts	2	0	Jan-Feb-Mar	2800
British Isles, North sea coasts, Scandinavia	2	3	April-May-June	3500
Italy, west coast around Pisa	2	9	Nov	4000
Egypt, River Nile	3	0	Jan-Feb-Mar	4500

TABLE 18: Time needed to grow elvers to 200 gm market size eels in natural temperature outdoor, uncovered still-water ponds in various areas of the world

Place	No. of months 23°C and above	Annual temp. range	Approx. temperatures °C of still-water ponds based on mean monthly values month-degrees annual total	Time needed for eels to grow to 200 gm
Germany (Berlin)	1	0-23	150	About 4 years
England (Somerset)	0	4-22	160	About 4 years
France (Marseilles)	2	7-27	200	About 3 years
New Zealand (Auckland)	2	12-25	200	About 3 years
Australia (Melbourne)	2	8-27	200	About 3 years
Japan (Hamanako)	4	5-30	220	About 2 years
Tunisia	3	10-30	220	About 2 years
Australia (Brisbane)	6	14-31	260	About 1 year
Taiwan	6	13-31	260	About 1.5 years
Indonesia	12	25-31	330	About 1 year

TABLE 19
Ratio of values of various Japanese Fisheries 1971

Sea Fisheries	Percent	
Tuna, Albacore	11.4	
Alaska pollack	5.5	
Salmon, Trout	4.0	
Flounder, Halibut	3.4	
Skipjack, etc.	3.0	
Mackerel	2.9	
Jack mackerel	2.5	
Sea bream	1.9	
Marlin, etc.	1.8	
Yellowtail	1.8	
Others	20.4	
Mollusca	14.3	
Crustacea	4.0	
Mammals	1.0	
Seaweed	2.0	79.9
Whales	2.8	2.8
Sea Culture		
Nori seaweed	5.9	
Yellowtail	3.1	
Pearls	1.0	
Others	2.3	12.3
Inland Water Fisheries	1.9	1.9
Inland Water Culture		
Eels	1.7	
Others	1.4	3.1
		100.0

Special Note

The illustrations and their captions that follow deserve to be studied with special care. They constitute a most valuable section of this book.

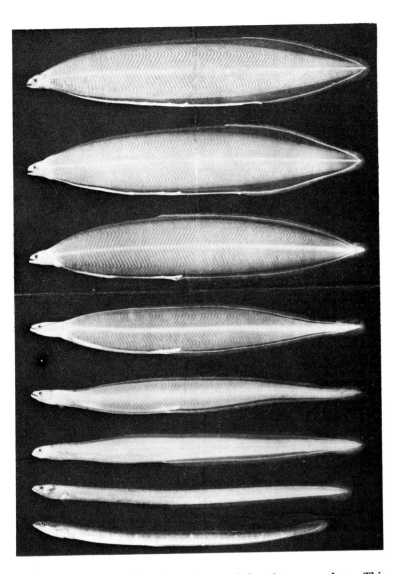

Fig 2 Metamorphosis of a leptocephalus into an elver. This illustration originally published in 1909 by the famous Johannes Schmidt is historic. It established the evolution of the eel from larvae which, when alive, are completely transparent. The elver at the bottom represents the final transformation. Thanks are due to the Danish Carlsberg Foundation for permission to republish this historic illustration.

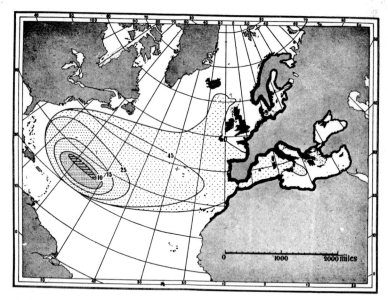

Fig 3   Distribution of the European eel [Anguilla anguilla] The figure
shows the end product of Johannes Schmidt's years of research which
proved that eels spawned in the sea and were originally marine fishes. The
only spawning ground for the European and the American eels is in the
Sargasso area of the tropical Atlantic Ocean. The Gulf Stream carries
leptocephalus larvae across to Europe and up the American coast. The
figures on the map show their length in millimetres at each stage of their
journey. The edge of the dotted zone marks the point at which
leptocephalae metamorphose into elvers. Note how leptocephalus
penetrate the Mediterranean. (after Schmidt 1922)

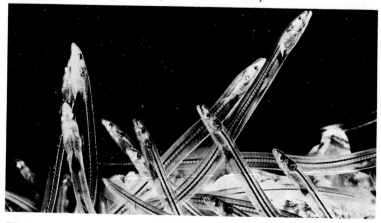

Fig 4   Transparent elvers newly caught in the River Severn fresh from
the sea. This is the starting point of eel culture. These elvers are already
some 2¼ years old. *Photo Williamson*

128

Fig 5   Elvers overcome waterfalls by wriggling up the wet mossy borders. They climb actively only at night; in the day they hide in crevices. This photo was taken in Victoria, Australia and shows [A. australis] elvers.

Fig 7   Mottled eel species [A. marmorata]. *Photo Williamson*

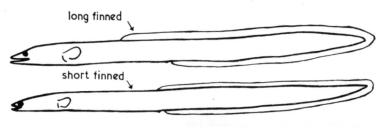

long finned

short finned

Fig 8   The dorsal fin of the 'long finned' species of eels starts well in front of the anus. In the 'short finned' species the dorsal and anal fin are more or less equal length.

Fig 9   Australian eel farmer weighs a consignment for market. [A. australis] differs only slightly from [A. japonica] and the [A. anguilla] the European eel, and is farmed in Tasmania & Victoria.

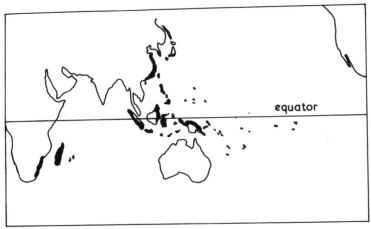

Fig 10   This map shows the distribution of [A. marmorata] marked in black. It is the most widely distributed species — its range spanning 12000 miles from South Africa to Japan and to the mid-Pacific Marquesas Islands.

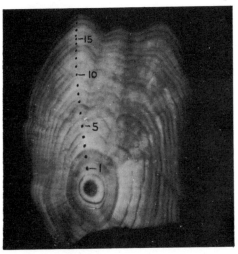

Fig 11   The otolith of an eel reveals its age. Otoliths are calcareous lumps that grow in fishes ears as part of the balancing apparatus. The rings reveal the growth of the fish like the annual rings in trees. This otolith came from an [A. rostrata] caught in Newfoundland, Canada. The innermost opaque white zone was formed at sea during the leptocephalus stage. Subsequent zones are marked with dots and were formed in freshwater during each summer's growth. This eel had spent one year in the sea as a larvae followed by 19 in freshwater, so it was 20 years old.

131

Fig 12   The use of fyke nets is one popular way of catching brown stage eels. *Photo Williamson*

Fig 13   Using baited eel pots is another successful method of catching brown stage eels. – *Photo Williamson*

Fig 14   Illustrated here is one way of catching silver stage eels at night on the River Bann, N Ireland. Such eels are most conveniently caught on migration into the sea in autumn. A permanent barrier of mesh or wattle can be constructed across the whole width of a river leaving openings in which nets can be set, the eels travel only at night and men must empty the net frequently. On one exceptional night in 1963 on the River Bann eight tons of eels were caught. In the River Po area in North Italy permanent weirs and installations provide the basis for a large eel processing industry which has been famous for centuries. – *Photo Williamson*

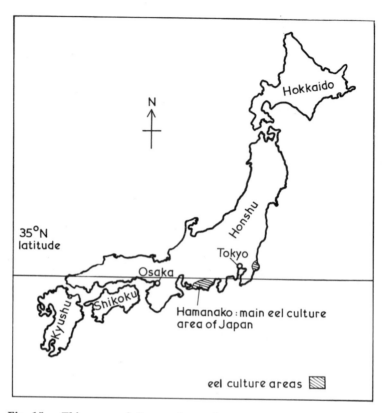

Fig 15 This map of Japan shows the areas where eel farming is carried out.

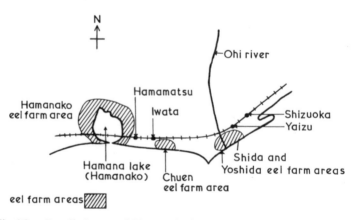

Fig 16 Detailed map of Hamanako lake area which is the main centre of eel culture in Japan.

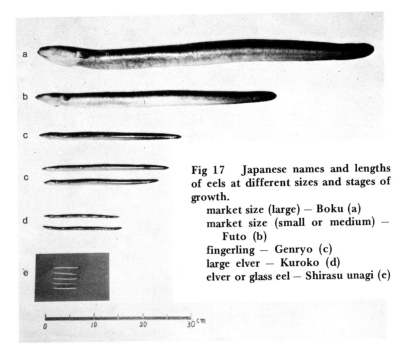

Fig 17   Japanese names and lengths of eels at different sizes and stages of growth.

market size (large) — Boku (a)
market size (small or medium) — Futo (b)
fingerling — Genryo (c)
large elver — Kuroko (d)
elver or glass eel — Shirasu unagi (e)

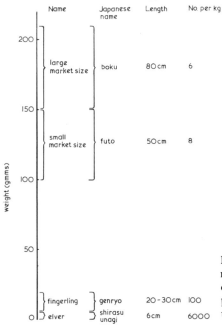

Name	Japanese name	Length	No. per kg
large market size	boku	80 cm	6
small market size	futo	50 cm	8
fingerling	genryo	20-30 cm	100
elver	shirasu unagi	6 cm	6000

Fig 18   Graph showing names and lengths of eels at different stages of growth. In practice small variations in the figures given may occur.

135

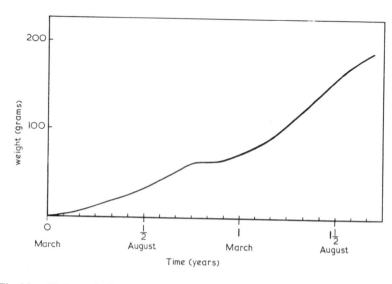

Fig 19   This graph shows growth rate of eels in a typical pond. Weight of 60gm is attained after one growing season, 190gm in two growing seasons. Both faster and slower rates commonly occur.

Fig 20   View of a typical eel farm in Japan with 1965 size ponds of about 0.5 hectare (a little over 1 acre)

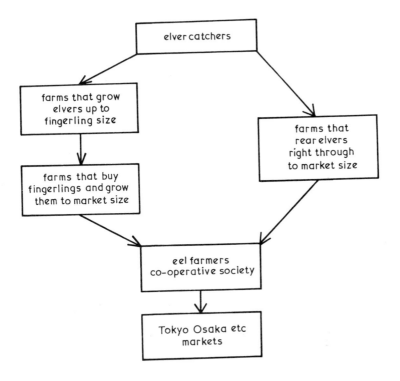

Fig 21 Organization of the eel culture industry in Japan. The establishment of the co-operatives is one of the principal reasons for the success of the industry.

Fig 22   Bulldozing flat land to make a pond.

Fig 23   A newly dug pond not yet lined.

Fig 24   One method of lining pond banks is the use of grooved concrete slats and posts.

Fig 25   A completed bank of grooved concrete slats and posts.

Fig 26   Preparing to build boulder and concrete pond banks. Note the drainage ditch outside the bank to the right.

Fig 27   Building a pond bank with boulders.

Fig 28   Empty pond showing mud bottom and solid concrete banks.

Fig 29   A solid concrete bank.

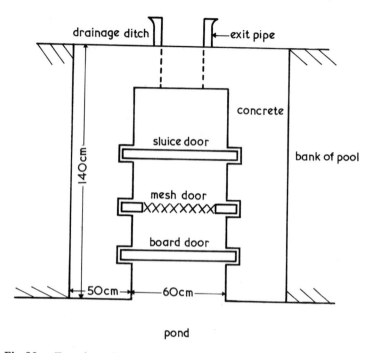

Fig 30 Top view of a sluice gate showing arrangement of board, mesh and sluice doors in their slots.

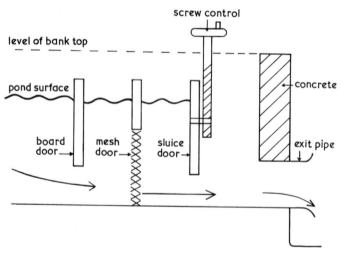

Fig 31 Vertical section of the sluice gate showing the three doors. The board door is set so that the water from the lower layers of the pond can be drained.

Fig 32   Building the exit sluice of a pond.

Fig 33   The completed exit sluice showing the three doors.

Fig 34   Drainage ditch lined with concreted boulders.

Fig 35   The exit sluice discharges into a drainage ditch by way of a large diameter concrete pipe. By tying a net bag over the lip of the pipe end, eels can be harvested by draining the pond.

Fig 36   Siphon arranged to drain water from the pond.

Fig 37   Pumps are used for raising water from a bore hole.

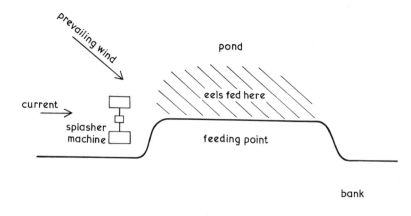

Fig 38   It is wise to establish a customary feeding point in the pond. This should be sited at a position where the oxygen level is high as this encourages the eels to eat well.

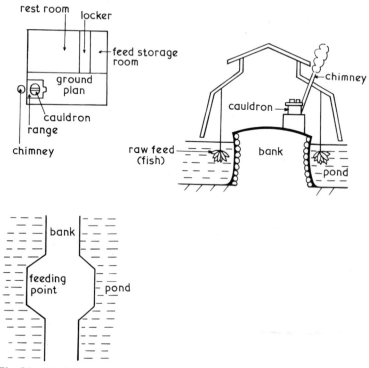

Fig 39   A hut and cooking equipment is sometimes provided at the feeding point.

Fig 40   A shaded feeding place.

Fig 41   Another type of a shaded feeding point.

147

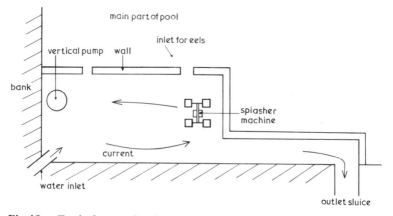

Fig 42   Typical example of a "resting corner" in an eel pond.

Fig 43   The "resting corner" in a pond.

Fig 44   Another design of a resting corner.

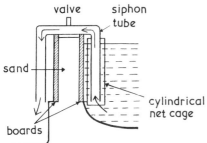

Fig 45   Siphons can be used for water outlets.

Fig 46   Here a siphon is being used to drain one pond into an adjacent empty pond. Pumps are for raising water up from a borehole.

149

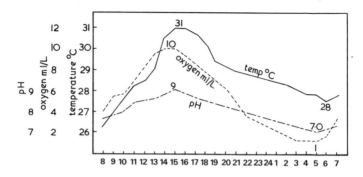

Fig 47　This diagram shows the normal daily cycle of water quality which occurs in a typical eel pond in summer. Temperature, oxygen, and pH scale, are at the side and the hours are indicated along the bottom.

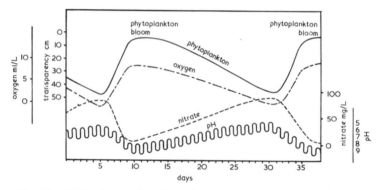

Fig 48　This shows the phytoplanktonic growth cycle and its effect on water quality. During the bloom the pond becomes a thick soupy green, nitrate is at a minimum and the pH is alkaline. According to the species of phytoplankton which grows in the pond and the weather, blooms occur every 40 days (approx.) during the summer.

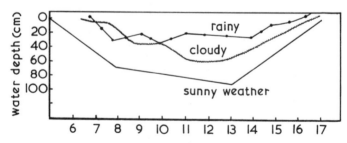

Fig 49　Graph shows depth at which depletion of oxygen must be compensated for by bringing in fresh water or increased by agitation in line with rain, cloud or sun (see gauge).

Fig 50   A typical paddle splasher machine for adding oxygen to the water.

Fig 51   A special paddle splasher having two pairs of paddles, one pair going deep and the other shallow.

Fig 52   A simple paddle splasher which is driven from the bank.

Fig 53   A vertical pump discharging onto a splash board to give added oxygenation.

Fig 54   A vertical pump in the resting corner of a pond.

Fig 55   A fountain aerator which is sometimes used is spectacular but
is not economical.

Fig 56   Artificial food for eels being mixed with water and vitamin oil. *Photo Williamson*

Fig 57   Eels ravenously eating mackerel. It is first dipped for a few minutes in boiling water to soften the skin.

Fig 58   Eels eating artificial food from a tray lowered into the pond. Once they have ceased feeding it is lifted out — this to avoid scraps fouling the water.

Fig 59   Suspended are shown the heads and backbones of mackerel which alone remain after the eels have fed. Threaded through the eyes these too are removed to avoid fouling the water.

Fig 60   A mincing machine for preparing the food.

Fig 61   A machine for mixing artificial food — the ingredients may vary slightly from time to time.

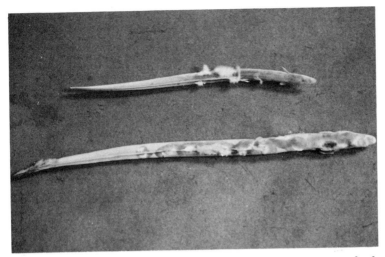

Fig 62   Fungus disease (Watamakuri-byo) Several hundred tons of eels were killed by this disease in Japan during 1966 and 1968.

Fig 63   Eels killed by fungus disease floating at the edge of a pond.

Fig 64   Sea gulls picking diseased eels from a pond.

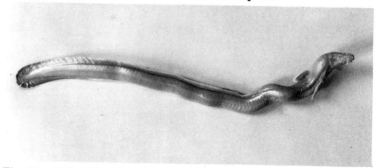

Fig 65   Crippled body disease (Beko-byo) is caused by protozoa attacking the muscles.

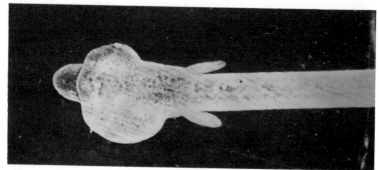

Fig 66   Gas disease (Kiho-byo) occurs in elvers.

158

Fig 67   White spot disease (Hakuten-byo) The white spots are colonies of the protozoan parasite (Myxidium).

Fig 68   Red disease (Hireaka-byo).

159

Fig 69   Gill disease (Era-byo)

Fig 70   Anchor worm disease (Ikarimushi-byo) Many parasitic anchor worms (Lernaea cyprinacea) attach themselves inside the mouth of the eel where they suck the blood and prevent the eel from feeding.

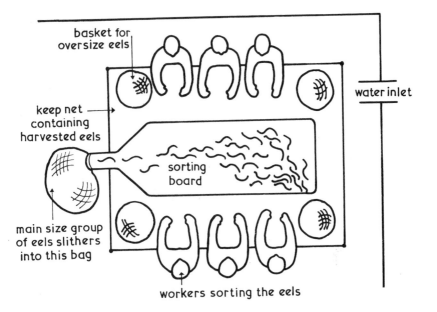

basket for oversize eels

keep net containing harvested eels

main size group of eels slithers into this bag

water inlet

sorting board

workers sorting the eels

Fig 71 An arrangement for sorting eels to different sizes. Eels to be harvested are held in a keep net staked in the resting corner of the pond. Eels are then scooped from the net onto the sorting table. Big eels are flicked into special baskets, and undersized eels are flicked back into the keep net to be grown further. The majority of the eels slither down the board into a bag net tied over the end.

Fig 72 Bag net tied over the exit pipe of a pond to harvest the eels by draining the pond.

Fig 73   Eels can be caught
by setting this net in front of
the exit sluice. The bamboo
curtain acts like the wing of a
seine to guide the eels into
the net.

Fig 74   Using a scoop net to catch eels. No food is given during the
previous two or three days and now as the eels come hungrily to the
food the men scoop them up.

162

Fig 75   A seine net can be used to harvest eels.

Fig 76   The seine net drawn tight. Care must always be taken to handle the eels carefully and avoid injuring the skin.

Fig 77   Netted eels are put into plastic baskets to be lifted on to the sorting table.

Fig 78   The keep net is placed in a corner of the pond close to the water inlet where the oxygen is at its maximum.

Fig 79   Eels being transferred from the seine net to the keep net.

Fig 80   A scene at the keep net.

Fig 81   The harvested eels are sorted and the small ones returned to the pond for further growth.

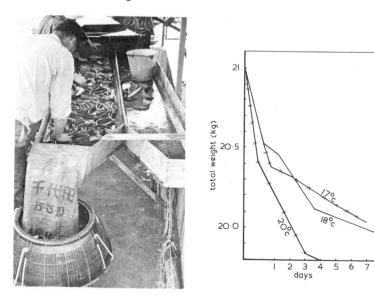

Fig 82   Another type of sorting table.

Fig 83   A weight loss of about five per cent occurs during starvation cleansing. This graph shows the rate of weight loss during starvation cleansing at temperatures of 17, 18 and 20°C.

166

Fig 84   Harvested eels are kept three days in running water without food before dispatch to market. This process, called "Ikeshime" allows the eels to complete digestion of food in their stomachs and empty all excretions from their guts. They are held in 'Doman' baskets in front of the water inlet of the pond.

Fig 85   Doman baskets of eels left for starvation cleansing in mid-pond.

167

Fig 86   The Doman storage baskets are made of polyethylene.

Fig 87   Sometimes the starvation cleansing of the eels is done by stacking the perforated plastic tubs under showers of trickling water.

Fig 88 Fingerling size eels are graded into different sizes by using mesh containers.

Fig 89 Here fingerlings in keep nets await sorting while a splasher aerator machine keeps up the oxygen supply.

Fig 90 10kg of eels is put in a double polyethylene bag inflated with oxygen.

Fig 91 Polyethylene bag inflated with oxygen is put in carton.

Fig 92 A tanker lorry used by a live-eel supply company in Britain. The compartments are aerated and can hold eels at densities of 30 lbs to the cubic foot 13°C or at 15 lbs to the cubic foot at 20°. In such lorries eels can stand transport on journeys of two weeks with the water being changed every four days. – *Photo Williamson*

Fig 93 A special barge with perforated sides and bottom stores 60 tons of eels at Maldon, England. The barge is kept in a canal but to prevent a certain disease is moved into sea water for two hours every two weeks – *Photo Williamson*

171

Fig 94 Eels being discharged from a modern tanker able to carry 15 tons of eels. Similar lorries carry cargoes of live eels from USA to Europe, the aerators working during the voyage. – *Photo Williamson*

Fig 95 Dip net used for catching elvers in the river Severn, England. Elvers sweep in on high tide and after the tide ebbs they start moving upstream. At promontories they hug the banks and best catches are made there. Sometimes a single dip of the net will secure 10kg of elvers. – *Photo Williamson*

172

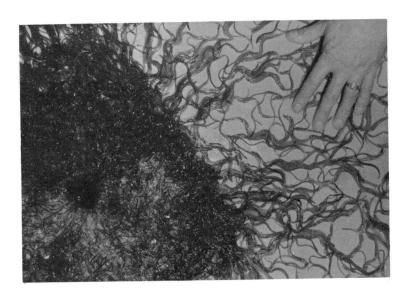

Fig 96 Newly caught elvers. — *Photo Williamson*

Fig 97 An elver catcher weighs his catch at the Epney-on-Severn collecting station. Of some 50 tons caught in this river yearly about half are eaten, a quarter exported to Germany and a quarter flown to Japan. — *Photo Williamson*

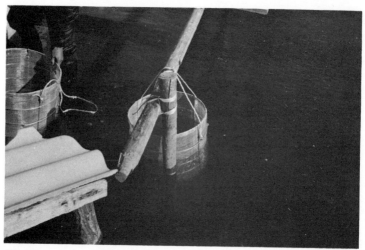

Fig 98   Gauze cages used by Japanese elver catchers for holding their catches. As elsewhere this involves great activity at the time but the business has expanded so much in Japan that more elvers are needed than can be supplied locally.

Fig 99   Tanks for holding elvers at Epney-on-Severn. — *Photo Williamson*

Fig 100   Emptying an elver holding tank. Note the square of perforated aeration hose at the bottom. — *Photo Williamson*

Fig 101   Loading elvers from the Severn into a tanker lorry to go to Hamburg, Germany. White froth generated by the elvers covers the ground. The lorry has six tanks, an air compressor to work bubble aerators and carries 1.2 tons of elvers. Salt is added to the tank to make the water half saline. The journey takes 15 hours. *Photo Williamson*

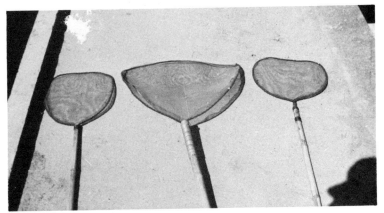

Fig 102   Japanese gauze scoops for catching elvers. The net part consists of smooth surfaced metal or gauze.

175

Fig 103   Enlarged ovaries of female eels that have been injected with hormones. So far, no one has succeeded in inducing captive eels to breed. Hence eel culture depends on the capture of enough wild elvers each year.

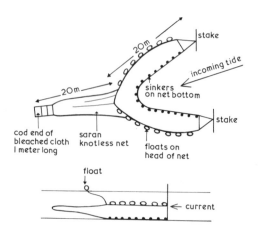

Fig 104   Net scheme for catching elvers. It is staked across a river with a float attached to the headline to keep the mouth wide open.

Fig 105   Latest type of trays used in air-freighting elvers from Europe to Japan. Each 0.5 kg heavy tray holds 1 kg of elvers. No ice is used; the elvers are simply cooled to 6°C before packing, put in the trays with only enough water to keep them wet and loaded on the plane. Millions and millions of elvers cross the North Pole like this every spring. — *Photo Williamson*

Fig 106   An older type of expanded polystyrene tray for air-freighting elvers. Modern type is cheaper and simpler. — *Photo Williamson*

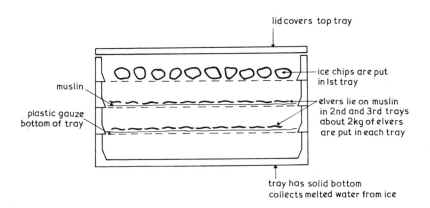

177

Fig 107   Kabayaki preparation. The eel is split open and the back-bone removed.

Fig 108   Bamboo skewers are put through the pieces of eel to keep them spread flat during cooking.

Fig 109   The eel, the colour of which first turns white by broiling, is then steamed and broiled a second time. (Eels are taken from the steam box.)

178

Fig 110   During broiling the eel is dipped into 'tare' sauce two or three times.

Fig 111   The pieces are grilled both sides over a charcoal fire to produce the finished Kabayaki.

Fig 112   When traditionally served a kabayaki meal is laid out on a low table on totami mats.

179

Fig 113 Australian smoked eels. The eels are de-slimed, gutted, dipped in brine and then smoked.

Fig 114 Jellied eels is the traditional British way of eating eels. About 800 tons of eels are so eaten in Britain each year and nearly all are sold at street stalls in London's East End. — *Photo Williamson*

Fig 115   Multi-span polyethylene greenhouses will possibly be used in
temperate countries to warm the water. Illustrated are some designed at
the Lee Valley Experimental Horticulture station in England.

Index

Pages, Tables and *Figures* are listed in this order

184

FISH FARMING

To provide the expanding fish farming activities of the world with a regular platform and central source of information there was established in May 1972 an outstanding journal under the title

Fish Farming International I

With its 150 pages packed with solid articles ranging over a wide field it was warmly welcomed the world over. Editor Peter Hjul, drawing on the resources of the journals *Fishing News* (weekly) and *Fishing News International* (monthly) organised an outstanding team of contributors in both scientific and practical circles. That issue was priced at £3 and gave excellent value for money.

A second volume, Fish Farming International 2 (same size, format and price) will appear in April 1973 with an even better array of valuable editorial content. Prominent in it will be a major article on the successful Scottish venture of breeding salmon without free range, as well as several articles by Dr. P H Milne (author of *Fish and Shellfish Farming In Coastal Waters*) arising from his recent visit to the fish farming areas of Japan and the Far East.

Succeeding volumes will be numbered numerically until a regular periodicity of half yearly or quarterly issue can be arranged — This will be done as soon as contributions can be assured in volume and quality for our purpose; because we are determined this work shall always be authoritative and worthy of the future we foresee for fish farming as a service to meeting human needs:

Make sure of a complete file of this great service publication by writing now for the issues you need. The price will be £3 per issue, post free, till a regular subscription basis can be assured.

Fishing News (Books) Ltd
23 Rosemount Avenue,
West Byfleet,
Surrey, England.

Books on Aquaculture

COASTAL AQUACULTURE IN THE INDO-PACIFIC REGION
Over 50 papers survey current and prospective development in Australia, Japan, Korea, Malaysia, Philippines, Taiwan, Thailand, USA, Vietnam, India, Ceylon, New Zealand. Edited T V R Pillay (FAO) 500 pages give research, experiments, practices, breeding, culture, marketing. £8.00

TEXTBOOK OF FISH CULTURE: Breeding and Cultivation of Fish
Marcel Huet's lifework. Magnificent work, covers every aspect of fish culture. 16 chapters, over 500 illustrations, 124 species of fish now cultured. Bibliography lists over 360 titles. £12.50

EUROPEAN INLAND WATER FISH
Multilingual catalogue lists 393 fish giving scientific and common names in languages of countries inhabited. Large carefully drawn figures supplemented by 25 colour plates. Key location maps of habitat. Four languages for main text English, German, French, Spanish. £8.75

FARMING THE EDGE OF THE SEA
From 1968 this work has pioneered progressive development in many countries. Six main sections treat present, future prospects, procedures, foods, feeding practices and fertilisers. Glossary, check list of essentials. Well illustrated £4.25

FISH AND SHELLFISH FARMING IN COASTAL WATERS
Most recent up-to-date work by engineering consultant and university lecturer Dr. P H Milne (Glasgow). Comprehensive and detailed. Covers site selection, design and construction of plants, full practical procedures. 208 pages, 151 modern figures, photographs. £7.50

FISHING WITH ELECTRICITY
FAO organised European Inland Fisheries Advisory Council experts to compile this masterly survey to serve for a decade as the last word on the subject. Fully detailed with practical guidance. 304 pages, 93 figures and 17 tables. £3.75

BRITISH FRESH WATER FISHES
Author-biologist Margaret Varley traces origin of fish fauna and outlines factors involved in choice of habitat giving useful guidance for sportsmen and culturists. 20 superb plates, 25 line figures.   £1.80

Other works are in preparation for early publication on ecology, environment, cultural practices, and the practical recognition and treatment of diseases of trout and salmon as well as the general diseases of fish.

These works, as they appear will be listed and noticed in the columns of our periodicals *Fishing News* and *Fishing News International,* the outstanding monthly of world-wide circulation.

In addition a catalogue giving the full list of available books is issued regularly and a copy will be sent free on request to all enquirers.

Further the newly launched publication *Fish Farming International* (see details elsewhere in these pages) regularly gives up-to-date information on all types of aquaculture.

Fishing News (Books) Ltd
23 Rosemount Avenue,
West Byfleet,
Surrey, England.